Intermediate Python and Large Language Models

Dilyan Grigorov

Apress®

Intermediate Python and Large Language Models

Dilyan Grigorov
Varna, Varna, Bulgaria

ISBN-13 (pbk): 979-8-8688-1474-7 ISBN-13 (electronic): 979-8-8688-1475-4
https://doi.org/10.1007/979-8-8688-1475-4

Managing Director, Apress Media LLC: Welmoed Spahr
Acquisitions Editor: Celestin Suresh John
Development Editor: Laura Berendson
Coordinating Editor: Gryffin Winkler

Cover image by Trevor M@Pixabay.com

Distributed to the book trade worldwide by Springer Science+Business Media New York, 1 New York Plaza, New York, NY 10004. Phone 1-800-SPRINGER, fax (201) 348-4505, e-mail orders-ny@springer-sbm.com, or visit www.springeronline.com. Apress Media, LLC is a Delaware LLC and the sole member (owner) is Springer Science + Business Media Finance Inc (SSBM Finance Inc). SSBM Finance Inc is a **Delaware** corporation.

For information on translations, please e-mail booktranslations@springernature.com; for reprint, paperback, or audio rights, please e-mail bookpermissions@springernature.com.

Apress titles may be purchased in bulk for academic, corporate, or promotional use. eBook versions and licenses are also available for most titles. For more information, reference our Print and eBook Bulk Sales web page at http://www.apress.com/bulk-sales.

Any source code or other supplementary material referenced by the author in this book is available to readers on GitHub (https://github.com/Apress). For more detailed information, please visit https://www.apress.com/gp/services/source-code.

If disposing of this product, please recycle the paper

To my son, my family, and the entire Python Software Foundation and AI community around the world.

Table of Contents

About the Author

Dilyan Grigorov is a software developer with a passion for Python software development, generative deep learning and machine learning, data structures, and algorithms. He was a Stanford student in the Graduate Program on Artificial Intelligence in the classes of people like Andrew Ng, Fei-Fei Li, and Christopher Manning. He has been mentored by software engineers and AI experts from Google and NVIDIA. Dilyan is an advocate for open source and the Python language itself. He has 16 years of industry experience programming in Python and has spent 5 of those years researching and testing generative AI solutions. His passion for them stems from his background as an SEO specialist dealing with search engine algorithms daily. He enjoys engaging with the software community, often giving talks at local meetups and larger conferences. In his spare time, he enjoys reading books, hiking in the mountains, taking long walks, playing with his son, and playing the piano.

About the Technical Reviewer

Tuhin Sharma is Sr. Principal Data Scientist at Red Hat in the Data & AI group. Prior to that, he worked at Hypersonix as an AI architect. He also cofounded and has been CEO of Binaize, a website conversion intelligence product for ecommerce SMBs. Previously, he was part of IBM Watson where he worked on NLP and ML projects, a few of which were featured on Star Sports and CNN-IBN. He received a master's degree from IIT Roorkee and a bachelor's degree from IIEST Shibpur in Computer Science. He loves to code and collaborate on open source projects. He is one of the top 25 contributors of pandas. He has four research papers and five patents in the fields of AI and NLP. He is a reviewer of the IEEE MASS conference, Springer Nature, and Packt publications in the AI track. He writes deep learning articles for O'Reilly in collaboration with the AWS MXNET team. He is a regular speaker at prominent AI conferences like O'Reilly's Strata Data & AI, PyCon, PyData, ODSC, GIDS, DevConf, etc.

Acknowledgments

I give a big thanks to the entire team at my publisher, Apress, and heartfelt thanks to two people who supported me throughout this book's writing process—Alexandre Blanchet (a software engineer with more than ten years of professional experience) and Haiguang Li. Alexandre's words about the book deeply moved me, and I'd love to share them with you:

> *I grew up in a small French city where speaking multiple languages was rare. From an early age, I loved computers and creativity, but wasn't sure about my career path. Everything changed in high school when I discovered coding—I had found my place.*
>
> *Back then, Python wasn't as popular, but I saw its potential. Despite limited job opportunities in my area, I committed to mastering it, believing in its future. Over the past decade, working at companies big and small has deepened my skills beyond what I learned in school.*
>
> *Today, I use that experience to guide students in accelerating their Python learning. That's how I met Dilyan. From the start, his curiosity and drive stood out. He dedicates himself fully to mastering his craft, and it has been a pleasure to support his journey.*
>
> *Dilyan quickly grasped Python and machine learning, leading to this book's creation. Beyond technical skills, he brings rare qualities—attention to detail, excellent time management, and versatility in marketing, SEO, business, and writing.*
>
> *Some people drain your energy; others inspire you. Dilyan is the latter. Our weekly coding sessions are always energizing, pushing us both to grow. I'm proud of this book—a testament to Dilyan's hard work, passion, and dedication. I'm confident it will inspire and empower your learning journey.*

Introduction

The evolution of artificial intelligence (AI) has ushered in a new era of possibilities, transforming the way we interact with technology, automate tasks, and solve complex problems. At the heart of this revolution are large language models (LLMs), which power applications ranging from conversational agents to content generation, data retrieval, and beyond. This book serves as an advanced comprehensive guide to understanding, developing, and deploying LLM-powered applications, with an advanced focus on Python and LangChain. It is designed for AI enthusiasts, data scientists, machine learning engineers, developers, and researchers who are looking to deepen their understanding of LLMs and their real-world applications.

The book bridges the gap between theory and practice, providing a road map for building advanced intelligent systems that leverage the power of language models. Throughout the chapters, I emphasize hands-on learning, providing code examples, best practices, and troubleshooting strategies to help you build efficient and effective AI-driven applications. By the end of your journey, you will have a strong foundation in LLMs and the ability to apply them to a wide range of real-world challenges.

The book is divided into several key chapters, each focusing on a critical aspect of working with LLMs and LangChain.

- **Chapter 1—LangChain and Python: Basics:** This chapter introduces the fundamentals of LangChain, a powerful framework for integrating LLMs into applications. It covers the core advanced concepts, including chains, memory, tools, and agents, along with how to structure prompts effectively for different tasks.
- **Chapter 2—LangChain and Python: Advanced Components:** Building on the basics, this chapter explores LangChain's advanced features, such as memory management, multiagent systems, and external data integrations. Readers will learn to create applications with contextual awareness and adaptability.

- **Chapter 3—Building Advanced Applications Powered by LLMs with LangChain and Python:** This chapter delves into the development of practical applications using LangChain and Python. It includes real-world examples like YouTube video summarizers and document retrieval tools, demonstrating how to implement advanced workflows and optimize model performance.
- **Chapter 4—Deploying LLM-Powered Applications:** Once an LLM application is built, deploying it effectively is crucial. This chapter covers cloud deployment strategies, model-serving solutions, optimization techniques, and best practices for ensuring scalability, security, and performance in production environments.
- **Chapter 5—Building and Fine-Tuning LLMs:** For those looking to take customization further, this chapter explains the principles of training and fine-tuning LLMs. It discusses transformer architectures, pretraining paradigms, fine-tuning strategies, and ethical considerations in deploying AI responsibly.

Happy reading and coding!

CHAPTER 1

LangChain and Python: Basics

LangChain is a powerful new framework in Python that simplifies building intelligent applications using natural language processing (NLP) and large language models (LLMs). It reduces complexity, making AI-powered solutions more accessible to developers. At its core, LangChain provides a set of abstractions and utilities that make it easier to build, customize, and deploy NLP-based workflows, such as chatbots, automated data analysis, summarization tools, and much more. Given Python's status as the go-to language for AI and data science, integrating LangChain with Python creates a powerful toolset for developers and data practitioners looking to enhance their NLP projects.

LangChain's primary goal is to simplify how developers interact with language models and manage their outputs in context-rich applications. Typically, when using a language model like OpenAI's GPT, there's a need to set up workflows for input, processing, context handling, and response generation. LangChain provides a framework to define and chain these elements, known as "chains," enabling more complex and sophisticated NLP applications without needing to manually handle all aspects of the process.

Python has a rich set of libraries for machine learning (e.g., TensorFlow, PyTorch) and NLP (e.g., spaCy, NLTK). LangChain seamlessly fits into this ecosystem by offering high-level abstractions that allow developers to quickly integrate language models into their applications. Key benefits include

- **Ease of Integration:** LangChain abstracts much of the complexity involved in setting up prompts, model calls, and response handling, making it easier to build and deploy applications.

D. Grigorov, *Intermediate Python and Large Language Models*, https://doi.org/10.1007/979-8-8688-1475-4_1

- **Modularity and Flexibility:** LangChain enables chaining multiple LLM calls together, combining different models, and adding context to create more advanced applications, such as multistep question-answering systems or conversational agents.
- **Handling Context and Memory:** One of LangChain's strengths is its ability to manage context and memory effectively. For conversational AI or tasks that require understanding of a sequence of interactions, LangChain provides utilities to track and store context throughout the conversation or workflow.
- **Scalability and Deployment:** By working within the Python ecosystem, LangChain can be easily integrated into larger projects, data pipelines, or cloud-deployed applications, making it a practical choice for both experimentation and production-level applications.

With Python and LangChain, developers can build a wide range of NLP applications:

- **Chatbots and Conversational Agents:** Implement agents that can handle context-aware conversations, manage user intents, and respond dynamically to user queries
- **Data Extraction and Summarization:** Create pipelines that process large amounts of text, extract key information, and produce summaries or insights
- **Automated Content Generation:** Use language models to generate content for blogs, reports, or documentation based on given prompts or templates
- **Question-Answering Systems:** Build tools that allow users to ask questions about specific documents or datasets, where the system can pull and present relevant information

In this first chapter, we will explore how to use LangChain with Python to create advanced language model applications, discussing its key components and providing practical examples to get you started.

This chapter also

- Introduces LangChain as a Python framework for building LLM-powered apps like chatbots and summarizers, with a focus on modularity, memory, and context handling
- Covers core concepts: chains, prompts, memory, tools, agents, RAG, data loaders, and integrations
- Explains installation of LangChain and related packages (e.g., langchain-core, langchain-openai, langgraph)
- Provides prompt engineering techniques: role prompting, few-shot, chain prompting, chain-of-thought, alternating messages, and refinement tips
- Describes various chain types: simple, sequential, conversational, multi-input/output, router, control flow, retrieval-aware, and agent chains—with code examples

LangChain Basics and Basic Components

As I mentioned, LangChain is a powerful framework designed for developing applications that integrate large language models (LLMs) like OpenAI's GPT-4 into workflows or pipelines that can perform a variety of complex tasks. It is particularly helpful for creating applications that require language model capabilities, whether for natural language understanding, processing, or generation.

Here are the fundamental components and concepts of LangChain.

Chains

Chains are sequences of operations (or steps) designed to process and transform data. In LangChain, chains can be created to link together multiple steps that involve LLMs, transforming the input through a sequence of transformations or tasks. A simple chain might involve querying an LLM with a prompt, whereas more complex chains can combine multiple actions, like API calls, data retrieval, or conditional logic.

Prompts

Prompts are the input that LLMs use to generate responses. LangChain allows users to design prompts dynamically, enabling the creation of tailored queries based on different scenarios or contexts. You can create prompt templates that include variables to be filled in based on user inputs or other data.

Memory

Memory allows a chain to retain state throughout a conversation or across multiple interactions. This feature is particularly useful for applications that require context over time, such as chatbots or assistants, where responses need to be informed by the history of the conversation.

Tools and Agents

LangChain provides tools that interact with external systems or APIs, such as databases, search engines, or custom APIs. Agents are advanced chains that can decide which tool to use based on the input they receive. For example, an agent could determine whether to perform a search, fetch data from a database, or generate a response directly.

Retrieval-Augmented Generation (RAG)

RAG is a method where LLMs are combined with external data sources to enhance their outputs. Unlike standard LLM queries that rely on pretrained knowledge, RAG dynamically retrieves up-to-date information from external sources before responding, ensuring better accuracy and contextual awareness. For example, an LLM may query a knowledge base or a search engine to find relevant information before generating a response. LangChain supports RAG through its retrieval tools and agents, making it suitable for applications that require updated or domain-specific information.

Data Loaders

LangChain includes data loaders for various types of data sources, like local files, APIs, and databases. These loaders help convert raw data into a format that can be processed or queried by an LLM.

Integrations and Extensibility

LangChain is designed to integrate easily with other tools and libraries. It supports various LLM back ends (such as OpenAI, Hugging Face, and others) and can be extended with custom chains, agents, or tools. This makes it flexible for creating custom applications across different domains.

LLM Outputs and Postprocessing

LangChain provides ways to interpret and process the outputs from LLMs. Since LLMs may produce complex or unstructured data, LangChain includes components for parsing, formatting, and further transforming these outputs to be more usable for the application.

By leveraging these concepts, LangChain allows for building powerful, customizable LLM-powered applications efficiently.

LangChain Installation

The LangChain ecosystem is divided into multiple packages, allowing you to selectively install only the specific features or functionality you need.

To install the main langchain package, run on Python 3.11:

```
pip install langchain==0.3.20
```

Although this package serves as a good starting point for using LangChain, its real value lies in integrating with various model providers and datastores. The necessary dependencies for these integrations are not included by default and must be installed separately. The steps to do so are provided below.

The LangChain ecosystem consists of different packages designed for modular functionality, most of which rely on "langchain-core." This package includes base classes and abstractions, providing a foundation for the rest of the ecosystem. When installing any package, you don't need to explicitly install its dependencies like "langchain-core." However, if you need features from a specific version, you may do so, ensuring compatibility with other integrations.

Packages Overview

- **LangChain Core:** Contains essential abstractions and LangChain Expression Language (LCEL). Automatically installed with "langchain" or separately with

  ```
  pip install langchain-core==0.3.41
  ```

- **Integration Packages:** Packages like "langchain-openai" or "langchain-anthropic" offer support for specific integrations. The complete list of these integrations can be found under the "Partner libs" section in the API reference of the LangChain documentation. To install any of them, use

  ```
  pip install langchain-openai==0.3.7
  ```

  Integrations that haven't been split into their own packages are part of "langchain-community," installed via

  ```
  pip install langchain-community==0.3.19
  ```

- **Experimental Package:** "langchain-experimental" hosts research and experimental code. You can install it with

  ```
  pip install langchain-experimental==0.3.4
  ```

- **LangGraph:** A library designed for building stateful, multiactor applications with LLMs, which integrates seamlessly with LangChain but can be used independently:

  ```
  pip install langgraph==0.3.5
  ```

- **LangServe:** A tool to deploy LangChain runnables and chains as REST APIs. It is included with the LangChain CLI. If you need both client and server functionalities, install using

  ```
  pip install "langserve[all]"
  ```

  For just the client or server, use `"langserve[client]"` or `"langserve[server]"`.

- **LangChain CLI:** Useful for managing LangChain templates and LangServe projects:

  ```
  pip install langchain-cli==0.0.35
  ```

- **LangSmith SDK:** Installed automatically with "langchain" but does not depend on "langchain-core." It can be used separately if you're not using LangChain:

  ```
  pip install langsmith==0.3.12
  ```

Installing from Source

To install any package from the source, clone the LangChain repository, navigate to the specific package's directory (e.g., "PATH/TO/REPO/langchain/libs/{package}"), and run

```
pip install -e .
```

This allows for flexible and targeted functionality, letting you selectively integrate or develop with specific packages in the ecosystem.

How to Prompt?

When working with large language models (LLMs), **prompt engineering** becomes an essential skill. A well-crafted prompt can significantly enhance the quality of a model's output, even when using less powerful or open source models. By understanding how to shape inputs effectively, you can guide LLMs to produce accurate, context-appropriate responses. Throughout this module, we'll explore the art and science of prompt creation, enabling you to fully harness the power of your models and achieve the best results possible.

One of the primary focuses will be on writing tailored prompts to achieve specific tasks, such as generating responses in a certain format or adhering to stylistic guidelines. We'll also examine how few-shot prompts can allow a model to quickly learn new tasks and generalize to unseen scenarios. This technique is especially useful when you need customization with minimal data, as it provides an efficient way to adapt model behavior on the fly.

Prompt Engineering

Prompt engineering is an emerging field focused on developing and refining prompts for effective use of large language models (LLMs) across a variety of applications. The goal is to enhance how LLMs process, understand, and generate text, making prompt engineering essential for numerous NLP tasks. Crafting high-quality prompts can reveal both the potential and the boundaries of what LLMs can achieve, and a well-designed prompt can significantly improve the accuracy and relevance of the model's responses.

Throughout this lesson, you'll gain hands-on experience with practical examples, helping you understand the nuances of prompt quality. We'll explore how different prompts can lead to significantly different results, highlighting what makes a prompt "good" or "bad." By the end, you'll be equipped with techniques to create powerful prompts that enhance model performance, enabling it to provide contextually relevant, accurate, and insightful responses to any given task.

Role Prompting

Role prompting is a technique that asks an LLM to take on a specific role or persona, helping guide its response in line with a certain tone, style, or perspective. For example, you might prompt the model to act as a *"copywriter," "teacher," or "data analyst."* This provides the LLM with a frame of reference, shaping how it interprets and answers the prompt.

To work effectively with role prompting, follow these steps:

- **Define the Role Clearly:** Clearly specify the role in your prompt to set the context for the model. For example, you might write: "As a copywriter, craft catchy taglines for AWS services that grab attention." The model will interpret the role and respond accordingly, adopting the language and style of a copywriter.
- **Generate Output from the LLM:** Once the role is defined, use your prompt to produce an output. The model will use the role as guidance to tailor its response appropriately, focusing on the style, language, or structure that aligns with the defined role.
- **Iterative Refinement:** Analyze the output to see if it meets the desired criteria. If the results are not as expected, refine the prompt by being more specific about the role or the style of the response.

This iterative process is crucial for achieving high-quality outputs. For example, if the response as a "copywriter" lacks creativity, you might adjust the prompt to include specific instructions like "use a playful tone and focus on benefits."

By guiding the model's behavior through role prompting, you can influence how it understands the task and the perspective it adopts, making it a versatile technique for a wide range of applications. This strategy not only improves the quality of the responses but also enables you to adapt the model's outputs to fit the context of different tasks more effectively.

Note For the following example, please get your OpenAI API key here: `https://platform.openai.com/api-keys`.

Example:

```
from langchain_core.prompts.prompt import PromptTemplate
from langchain_openai import ChatOpenAI

# Initialize the LLM with OpenAI's model

llm = ChatOpenAI(api_key=os.getenv("OPENAI_API_KEY"), model_name="gpt-4",
temperature=0.5)
template = """
As a futuristic poet, I want to write a poem that captures the essence of
{emotion}.
Can you suggest a title for a poem about {emotion} set in the year {year}?
"""

prompt = PromptTemplate(
    input_variables=["emotion", "year"],
    template=template,
)

# Input data for the prompt
input_data = {"emotion": "solitude", "year": "2500"}

chain = prompt | llm

response = chain.invoke(input_data)
```

```
print("Emotion: solitude")
print("Year: 2500")
print("AI-generated poem title:", response)
```

Output:

```
Emotion: solitude
Year: 2500
AI-generated poem title: content='"Echoes in the Void: Solitude in the 26th
Century"' additional_kwargs={'refusal': None} response_metadata={'token_
usage': {'completion_tokens': 16, 'prompt_tokens': 44, 'total_tokens': 60,
'completion_tokens_details': {'audio_tokens': None, 'reasoning_tokens':
0}, 'prompt_tokens_details': {'audio_tokens': None, 'cached_tokens': 0}},
'model_name': 'gpt-4-0613', 'system_fingerprint': None, 'finish_reason':
'stop', 'logprobs': None} id='run-c41b514e-f8f5-43a9-96c3-f0ab35fdaad6-0'
usage_metadata={'input_tokens': 44, 'output_tokens': 16, 'total_tokens':
60, 'input_token_details': {'cache_read': 0}, 'output_token_details':
{'reasoning': 0}}
```

The prompt in this code is effective for several reasons:

1. **Clear and Contextual Role Setting**

   By stating, "As a futuristic poet," the prompt establishes a role and context. This framing helps guide the model to think creatively like a poet, shaping its response to reflect a poetic tone and futuristic theme. Such context allows the LLM to adopt the right style, making the output more imaginative and relevant.

2. **Specificity of Emotion and Time Frame**

   The prompt specifically asks for a poem title that captures the emotion of "{emotion}" set in the year "{year}." This precision helps the model generate contextually rich and emotionally relevant titles, directly related to the emotion and future scenario. The use of variables makes it adaptable for different contexts, creating versatility.

3. **Open-Ended Creativity**

   The prompt is open-ended, allowing the LLM to generate diverse, creative titles without being overly restrictive. By not setting limitations on how the title should sound, the model can explore artistic and evocative language, enhancing the quality of the output.

4. **Task-Focused Guidance**

   The primary task is to create a poem title that evokes a specific emotion in a futuristic context. This direct focus helps the LLM avoid unrelated content, concentrating only on creating a unique title that matches the theme and style outlined in the prompt.

5. **Encouragement of Thematic Coherence**

   By guiding the LLM to align its output with an emotional and futuristic time frame, the prompt ensures the response will have both thematic and temporal coherence. This makes the resulting poem title not just relevant but also compelling and imaginative, showcasing how prompts can evoke specific styles and tones effectively.

Few-Shot Prompting

Few-shot prompting is a technique used in the context of large-scale language models to guide the model's output by providing a small number of task-specific examples within the input prompt. Unlike traditional machine learning approaches, which require extensive datasets and iterative training, few-shot prompting leverages a model's pre-existing knowledge to perform tasks with minimal supervision.

In few-shot prompting, the model is presented with a limited number of input/output pairs—usually between one and five—that illustrate the desired task. These examples serve as a form of implicit training within the prompt itself. The model uses these pairs to infer the relationship between inputs and outputs, allowing it to generalize and respond appropriately to new, unseen queries that follow the same pattern.

This technique builds on the premise that large language models, trained on vast amounts of diverse text data, can generalize across different domains. By presenting a few examples, the model can adjust its behavior dynamically without the need for explicit retraining or fine-tuning. Few-shot prompting thus demonstrates the flexibility and contextual reasoning ability of such models, allowing them to perform a wide range of tasks from a minimal set of instructions.

The effectiveness of few-shot prompting depends largely on the model's capacity to understand and generalize from the examples provided. It is a powerful approach for tasks where extensive labeled data is not readily available, offering an efficient method for leveraging pretrained models in a variety of applications.

Key Benefits

- **No Additional Training**: You don't need to fine-tune the model; it can perform tasks based on the few examples given.
- **Adaptability**: It can handle multiple tasks by simply providing examples for different tasks.
- **Efficiency**: Fewer examples are needed compared to traditional training methods, making it a practical approach for many applications.

Few-shot prompting is especially effective with very large pretrained models like GPT-3, which have enough capacity to learn from minimal examples.

Example:

```
from langchain_core.prompts.few_shot import FewShotPromptTemplate
from langchain_core.prompts.prompt import PromptTemplate
from langchain_openai import ChatOpenAI

# Initialize the language model with specific settings
language_model = ChatOpenAI(
    api_key="sk-proj-O56py5goMfqp8_g2gOgfhefr1HLriyWyP6erQJ4dQyi3D2HWBxJgCW
    rjWMbvMTJdvxHlzaWm11T3BlbkFJss1mhhNZJ7YREWFugP2wKQoMHIR3FMCDZxiOA_rPSrC
    fXZK6ZJbcGJ85dpMGV4adCt7R_zrUkA",
    model_name="gpt-4o-mini",
    temperature=0
)
```

```
# Sample color-to-emotion associations
color_emotion_pairs = [
    {"color": "red", "emotion": "energy"},
    {"color": "blue", "emotion": "peace"},
    {"color": "green", "emotion": "growth"},
]

# Template for formatting examples in a structured way
example_structure = """
Color: {color}
Associated Emotion: {emotion}\n
"""

# Create the example prompt template
color_prompt_template = PromptTemplate(
    input_variables=["color", "emotion"],
    template=example_structure,
)

# Construct a few-shot prompt template using the color-emotion pairs
few_shot_color_prompt = FewShotPromptTemplate(
    examples=color_emotion_pairs,
    example_prompt=color_prompt_template,
    prefix="Here are a few examples demonstrating the emotions linked with
    colors:\n\n",
    suffix="\n\nNow, considering the new color, predict the associated
    emotion:\n\nColor: {input}\nEmotion:",
    input_variables=["input"],
    example_separator="\n",
)

# Generate the final prompt for a new color input
final_prompt_text = few_shot_color_prompt.format(input="purple")

# Use the generated prompt and run it through the language model
final_prompt = PromptTemplate(template=final_prompt_text, input_
variables=[])
prompt_chain = final_prompt | language_model
```

```
# Get the AI-generated response for the input color
model_output = prompt_chain.invoke({})

# Print the input color and its corresponding predicted emotion
print("Color: purple")
print("Predicted Emotion:", model_output.content)
```

Output:

```
Color: purple
Predicted Emotion: Color: purple
Associated Emotion: creativity
```

Alternating Human/AI Messages

This strategy involves using few-shot prompting with alternating human and AI responses. It's particularly useful for chat-based applications, as it helps the language model grasp the flow of conversation and generate contextually relevant replies.

Though this method excels in handling conversational dynamics and is simple to implement for chat applications, it is less adaptable for other types of use cases and works best with chat-specific models. However, alternating human and AI messages can be applied creatively, such as building a prompt to translate English into pirate language in a chat format.

Chain Prompting

Chain prompting is a technique where multiple prompts are linked together in a sequence, with the output of one prompt being used as the input for the next. This method allows for progressively refining or expanding the context of the interaction, enabling the model to handle more complex tasks or multistep reasoning.

Key Characteristics

1. **Sequential Flow**: The process involves feeding the output from one step directly into the next, enabling the model to "remember" and build upon previous information.

2. **Dynamic Adjustments**: At each step, new information can be introduced based on the model's prior responses, allowing for iterative improvements in the result.

Steps for Chain Prompting

1. **Initial Prompt**: Start by providing an initial prompt to generate a base response.
2. **Extract Information**: Identify relevant details or key elements from the generated output.
3. **New Prompt Construction**: Create a subsequent prompt using the extracted information, adding new context or instructions to refine the output further.
4. **Repeat Process**: Continue chaining prompts as necessary, each building on the last, until the desired final output is obtained.

Using Chain Prompting in LangChain

To implement chain prompting in LangChain, you can leverage its **`PromptTemplate`** class. This class simplifies the construction of prompts by allowing for dynamic input values, making it ideal for situations where prompts need to evolve based on previous answers.

- **PromptTemplate** enables you to
    - Build prompts that adapt dynamically to changing inputs, ensuring flexibility in prompt chains
    - Simplify the process of passing outputs from one step to the next by easily substituting variables or new context into each prompt

Additional Benefits

- **Complex Workflows**: Chain prompting allows for handling more advanced tasks that require multiple steps, such as multiturn conversations, solving multipart problems, or conducting research in stages.
- **Error Handling**: If an intermediate step yields an incomplete or ambiguous response, chain prompting enables you to adjust the following prompts to clarify or correct the issue.
- **Interactive Exploration**: This approach allows for a more exploratory dialogue, where each prompt can refine the context, helping to uncover deeper insights.

In LangChain, combining chain prompting with other techniques like **few-shot prompting** or **memory-based approaches** allows you to build complex, multistep systems that leverage the power of large language models effectively.

Example:

```
from langchain_core.prompts.prompt import PromptTemplate
from langchain_openai import ChatOpenAI

# Initialize the language model
llm = ChatOpenAI(api_key="sk-proj-056py5goMfqp8_g2g0gfhefr1HLriyWyP6erQJ4dQ
yi3D2HWBxJgCWrjWMbvMTJdvxHlzaWm11T3BlbkFJss1mhhNZJ7YREWFugP2wKQoMHIR3FMCDZx
iOA_rPSrCfXZK6ZJbcGJ85dpMGV4adCt7R_zrUkA",
                 model_name="gpt-4o-mini",
                 temperature=0)

# Prompt 1: Ask for the scientist who developed the theory of general
relativity
question_template = """Who is the scientist that formulated the theory of
general relativity?
Answer: """
prompt_for_scientist = PromptTemplate(template=question_template, input_
variables=[])
```

```
# Prompt 2: Ask for a brief explanation of the scientist's theory of
general relativity
fact_template = """Give a brief explanation of {scientist}'s theory of
general relativity.
Answer: """
prompt_for_fact = PromptTemplate(input_variables=["scientist"],
template=fact_template)

# Create a runnable chain for the first prompt to retrieve the
scientist's name
chain_for_question = prompt_for_scientist | llm

# Get the response for the first question
response_to_question = chain_for_question.invoke({})

# Extract the scientist's name from the response
scientist_name = response_to_question.content.strip()

# Create a runnable chain for the second prompt using the extracted
scientist's name
chain_for_fact = prompt_for_fact | llm

# Input data for the second prompt
fact_input = {"scientist": scientist_name}

# Get the response for the second question about the theory
response_to_fact = chain_for_fact.invoke(fact_input)

# Output the scientist's name and the explanation of their theory
print("Scientist:", scientist_name)
print("Theory Description:", response_to_fact)
```

Output:

```
Scientist: The scientist who formulated the theory of general relativity is
Albert Einstein.
Theory Description: content="Albert Einstein's theory of general
relativity, formulated in 1915, is a fundamental theory of gravitation
that describes gravity not as a force, but as a curvature of spacetime
caused by mass. According to this theory, massive objects like planets and
```

```
stars warp the fabric of spacetime around them, and this curvature affects
the motion of other objects, causing them to follow curved paths. General
relativity has profound implications for our understanding of the universe,
including the behavior of black holes, the expansion of the universe, and
the bending of light around massive objects. It has been confirmed through
numerous experiments and observations, making it a cornerstone of modern
physics." additional_kwargs={'refusal': None} response_metadata={'token_
usage': {'completion_tokens': 131, 'prompt_tokens': 36, 'total_tokens':
167, 'completion_tokens_details': {'audio_tokens': None, 'reasoning_
tokens': 0}, 'prompt_tokens_details': {'audio_tokens': None, 'cached_
tokens': 0}}, 'model_name': 'gpt-4o-mini-2024-07-18', 'system_fingerprint':
'fp_e2bde53e6e', 'finish_reason': 'stop', 'logprobs': None} id='run-
d16897c9-54a3-4feb-9a7c-fe481798c984-0' usage_metadata={'input_tokens': 36,
'output_tokens': 131, 'total_tokens': 167, 'input_token_details': {'cache_
read': 0}, 'output_token_details': {'reasoning': 0}}
```

Chain-of-Thought Prompting

Chain-of-thought prompting (CoT) is a technique designed to encourage large language models (LLMs) to explain their reasoning process, leading to more accurate outcomes. By presenting few-shot examples that showcase step-by-step reasoning, CoT helps guide the model to articulate its thought process when responding to prompts. This method has proven effective for tasks such as arithmetic, common sense reasoning, and symbolic logic.

In the context of LangChain, CoT offers several advantages. First, it helps deconstruct complex problems by guiding the model to break them into smaller, more manageable steps, which makes the problem easier to solve. This is especially useful for tasks involving calculations, logic, or multistep reasoning. Second, CoT can help the model generate more coherent and contextually relevant outputs by leading it through related prompts. This results in more accurate and meaningful responses, particularly for tasks that require deep comprehension of a problem or domain.

However, there are some limitations to CoT. One significant drawback is that it generally improves performance only when applied to models with approximately 100 billion parameters or more. Smaller models often generate illogical reasoning chains, which can result in lower accuracy compared to standard prompting. Additionally,

CoT's effectiveness varies across different types of tasks. While it excels in tasks involving arithmetic, common sense, and symbolic reasoning, it may offer fewer benefits or even hinder performance in other task categories.

Advanced Tips for Effective Prompt Engineering

1. **Be Specific with Your Prompt**: Provide clear and detailed instructions in your prompt. The more context, background, and specifics you give, the better the LLM can interpret and generate a relevant response. Vague prompts lead to generalized or incomplete answers.

2. **Encourage Conciseness**: If the response needs to be short and to the point, be explicit about it. You can request responses to be limited to a specific number of words or sentences, which forces the model to focus on delivering the essential information.

3. **Ask for Reasoning or Explanations**: When dealing with complex tasks, encourage the model to explain its reasoning or show the steps it took to arrive at its answer. This improves the quality of results, particularly for problem-solving, logic, and reasoning tasks, ensuring transparency in the process.

4. **Iterate and Refine Prompts**: Prompt engineering is rarely a one-time activity. Iteration is key—test and tweak your prompts to see how different phrasing or added details change the model's response. Refine until the output aligns with your expectations.

5. **Use Examples to Guide Responses**: One of the most powerful ways to guide LLMs is by using few-shot learning. By showing the model a few examples of what you're looking for, you significantly increase the chance of receiving an answer that mirrors your expectations in tone, format, or reasoning.

6. **Apply Constraints**: If you're looking for specific formats or a particular structure (e.g., bulleted lists, headings, step-by-step processes), be clear about these constraints in your prompt. This helps the model organize its output according to your needs.

7. **Task-Specific Prompting**: Tailor your prompts to the specific task at hand. For example, creative writing prompts should encourage open-ended responses, while technical prompts should focus on precision, structure, and accuracy. Each type of task may require a different approach to prompt engineering.

8. **Leverage Clarifying Questions**: If the initial response isn't what you expected, ask the model to elaborate or clarify specific points. This helps guide the conversation in a more meaningful direction and ensures the model understands and focuses on what's important.

9. **Balance Open-Endedness and Constraints**: For tasks where creativity is needed, such as brainstorming, use more open-ended prompts to allow the model to explore a variety of ideas. For tasks requiring accuracy, use tighter constraints to keep the model focused on relevant and correct answers.

10. **Adjust Prompt Length**: The length of your prompt can influence the quality of the response. For some tasks, a simple, concise prompt works best, while more complex tasks might require detailed, multipart instructions. Experiment with prompt length to see what works for different types of questions.

11. **Include Key Terms**: If your task requires specific technical language, jargon, or domain-specific terms, include those directly in the prompt. This helps guide the model toward more specialized and accurate outputs, especially in fields like science, technology, or law.

12. **Specify the Role of the LLM**: Sometimes, framing the model's role in the prompt can improve the result. For instance, start your prompt with phrases like "As a teacher," or "You are an expert in..." to influence the model's tone and style of response, aligning it with the required task.

13. **Set an Output Persona**: In certain tasks, you can request the model to assume a specific persona or tone. For example, ask the model to respond like a teacher, researcher, or customer service agent to tailor the responses to different contexts or audiences.

14. **Utilize Multiturn Dialogue**: For tasks that require deeper exploration, consider breaking the problem down into a series of smaller questions. This approach not only helps the model focus on individual components of a complex task but also provides you with an opportunity to guide the conversation progressively toward a complete answer.

15. **Test Edge Cases**: For robustness, test how your prompt performs with edge cases or atypical inputs. This helps ensure that the LLM performs well across a variety of scenarios and doesn't generate inaccurate or nonsensical results in unusual situations.

16. **Account for Model Limitations**: Remember that LLMs have limitations in their knowledge and reasoning capabilities. Not all prompts will yield perfect responses, and some answers might lack depth or accuracy in certain specialized domains. Recognize when an LLM has reached its limit, and avoid overrelying on it for highly specialized or sensitive tasks.

17. **Keep Bias in Check**: Be mindful of the potential for biases in LLM-generated outputs. Craft prompts that minimize the chances of generating biased, harmful, or inappropriate content. Avoid phrasing that could steer the model toward biased or harmful assumptions.

18. **Incorporate Multiple Prompt Variations**: Instead of relying on one version of a prompt, try asking the same question or requesting the same task using several different prompt phrasings. This technique helps in uncovering new insights or variations in response quality.

By applying these strategies, you can enhance your ability to interact effectively with large language models, improving the quality and relevance of their outputs. As AI tools continue to evolve, mastery of prompt engineering will remain a critical skill for developers, researchers, and professionals who rely on LLMs to optimize their workflows.

What Are Chains?

A LangChain chain is a structured sequence of operations in the LangChain framework, where various components like language models, tools, and external APIs are connected to perform complex tasks. The primary purpose of a chain is to manage and coordinate interactions between different modules, allowing for multistep reasoning and advanced workflows when working with large language models (LLMs).

Key characteristics of a LangChain chain:

- **Modular Design:** Chains are designed to be modular, meaning individual components can be easily added, removed, or replaced. This allows for flexibility in constructing workflows depending on the use case, from simple to highly sophisticated tasks. Each module or component typically has a clearly defined input/output structure.
- **Multistep Processing:** Chains facilitate multistep operations by passing the output of one component as the input to another. This enables more advanced reasoning, decision-making, or actions that require several stages of processing, such as combining language understanding with tool execution or validation.
- **Control Flow:** Chains can incorporate control flow mechanisms, such as conditional logic or loops, enabling the workflow to branch or iterate based on the intermediate results. This allows for dynamic behavior, adjusting the sequence of actions depending on the inputs or outputs at each step.
- **Handling Intermediate Outputs:** A chain can retain intermediate outputs, either for logging purposes, debugging, or as part of a larger workflow. This allows for transparency in the process, making it easier to inspect how each step contributes to the final result.
- **Interaction with External Systems:** Chains are not limited to just working with language models. They can interact with external systems, such as databases, APIs, search engines, or knowledge bases, to fetch relevant information or execute tasks that go beyond natural language processing. This is particularly useful for retrieving real-time data, performing calculations, or executing functions that require interaction with other platforms.

- **Memory Management:** Some chains integrate memory, allowing them to store and recall past interactions, decisions, or context. This feature is particularly valuable for applications like conversational agents, where maintaining context over multiple interactions is critical for coherent and contextually aware responses.
- **Scalability:** Chains can be constructed in a scalable manner, allowing developers to design workflows that handle both simple tasks (such as a single prompt) or more intricate, multistep processes involving numerous components and external services.
- **Reusability:** LangChain encourages reusability by enabling the creation of reusable chains that can be applied to different tasks without reconfiguring the entire workflow. Developers can design a chain once and use it for various applications or modify it for similar tasks with minimal changes.

LangChain chains are an essential mechanism for building sophisticated applications that go beyond simple LLM queries, orchestrating complex interactions in a seamless, structured, and highly configurable way.

Chain Components

A LangChain chain consists of several key components that work together to create multistep workflows.

First, **prompt templates** are used to guide LLM outputs by filling in placeholders with dynamic values, helping customize the responses. The core of the system, **language models (LLMs)**, generate responses based on the input prompts. Chains can also integrate with **external tools**, such as APIs or databases, to fetch data or perform additional tasks beyond text generation.

Memory is another crucial component, allowing the chain to store and recall information across interactions, ensuring continuity, especially in conversational contexts. **Input variables** provide dynamic data that personalize the chain's behavior, while **output parsers** process and format model outputs for further steps or final responses.

More complex tasks can be handled by **nested chains (subchains)**, which break down workflows into smaller, manageable steps. **Decision logic** introduces conditional branching, enabling the chain to adapt based on input or intermediate results. Chains can also include **retrieval components** to fetch relevant information from external sources, enhancing context and accuracy.

Control flow governs the sequence and timing of operations, ensuring tasks are performed in the right order. To ensure robustness, **error handling mechanisms** are built in, managing failures and triggering retries or alternative steps when needed. **API connectors** allow chains to interact with external services, expanding functionality, while **logs and debugging** tools track execution, helping with monitoring and troubleshooting.

These components enable LangChain chains to integrate LLMs with tools, logic, and external data sources, allowing for flexible and complex workflows tailored to various applications.

Chain Types

In LangChain, there are several types of chains that can be used to construct workflows depending on the complexity, purpose, and specific requirements of the task. Each chain type serves a different function and can be adapted or combined to create versatile applications. Here are the most common types of **LangChain chains**.

1. Simple Chain

A simple chain consists of a single operation or a straightforward sequence of operations. This type of chain takes an input, processes it through one or more steps, and generates a single output. It's often used for basic tasks, such as filling in a prompt template and calling an LLM to generate a response.

- **Usage**: Direct question-answering tasks, summarization, or text transformation
- **Components**: Usually involves a single prompt template, one LLM call, and an output

Example:

```
from langchain.chains import LLMChain
from langchain.prompts import PromptTemplate

from langchain_openai import OpenAI

# Step 1: Define the language model (in this case, OpenAI's GPT)
llm = OpenAI(api_key="sk-proj-O56py5goMfqp8_g2gOgfhefr1HLriyWyP6erQJ4dQyi3D
2HWBxJgCWrjWMbvMTJdvxHlzaWm11T3BlbkFJss1mhhNZJ7YREWFugP2wKQoMHIR3FMCDZxiOA_
rPSrCfXZK6ZJbcGJ85dpMGV4adCt7R_zrUkA",
            temperature=0.7)  # Set the desired temperature for response
            variability

# Step 2: Define the prompt template
prompt_template = """
Summarize the following question briefly:
{user_question}
"""

# Step 3: Create the PromptTemplate object
prompt = PromptTemplate(
   input_variables=["user_question"],
   template=prompt_template,
)

# Step 4: Create the LLMChain using the language model and prompt template
 chain = prompt | llm

# Step 5: Input the user's question and run the chain
user_question = "Can you explain how photosynthesis works in simple terms?"

output = chain.invoke(user_question)

# Print the summarized question
print("Summarized Question:", output)
```

Output:

```
Summarized Question:
Explaining photosynthesis in simple terms.
```

2. Sequential Chain

A sequential chain involves multiple steps arranged in a strict linear sequence. Each step's output becomes the input for the next step. These chains are useful when tasks need to be completed in a particular order.

- **Usage**: When multistep reasoning or progressive tasks are needed (e.g., generating an outline, followed by writing content based on that outline)
- **Components**: Multiple operations, such as LLM calls, external API interactions, or data transformations that occur in sequence

Example:

```
from langchain.chains import LLMChain, SimpleSequentialChain
from langchain.prompts import PromptTemplate

from langchain_openai import OpenAI

# Step 1: Define the language model
llm = OpenAI(api_key="sk-proj-O56py5goMfqp8_g2gOgfhefr1HLriyWyP6erQJ4dQyi3D
2HWBxJgCWrjWMbvMTJdvxHlzaWm11T3BlbkFJss1mhhNZJ7YREWFugP2wKQoMHIR3FMCDZxiOA_
rPSrCfXZK6ZJbcGJ85dpMGV4adCt7R_zrUkA",
             temperature=0.7)

# Step 2: Create the first prompt template to summarize the question
summary_prompt_template = """
Summarize the following question briefly:
{user_question}
"""

# Step 3: Create the second prompt template to generate a short answer
answer_prompt_template = """
Provide a brief answer to the following question:
{summarized_question}
"""

# Step 4: Create PromptTemplate objects for both prompts
summary_prompt = PromptTemplate(
```

```
    input_variables=["user_question"],
    template=summary_prompt_template,
)
answer_prompt = PromptTemplate(
    input_variables=["summarized_question"],
    template=answer_prompt_template,
)

# Step 5: Create LLMChain objects for both steps
summary_chain = LLMChain(llm=llm, prompt=summary_prompt)
answer_chain = LLMChain(llm=llm, prompt=answer_prompt)

# Step 6: Create a SimpleSequentialChain that links both chains together
sequential_chain = SimpleSequentialChain(
    chains=[summary_chain, answer_chain]
)

# Step 7: Input the user's question and run the sequential chain
user_question = "Can you explain how photosynthesis works in simple terms?"
output = sequential_chain.run(user_question)
# Print the output of the sequential chain
print("Final Output:", output)
```

Output:

```
Photosynthesis is the process by which plants and some other organisms use
sunlight to turn water and carbon dioxide into oxygen and sugar. This sugar
is then used as a source of energy for the plant's growth and development.
The process takes place in the chloroplasts of plant cells and requires
the presence of chlorophyll, a green pigment that absorbs sunlight. During
photosynthesis, carbon dioxide is taken in through small openings on the
leaves called stomata, and water is absorbed through the roots. Sunlight
is then used to convert these substances into energy in the form of sugar,
while oxygen is released as a byproduct. This process is vital for the
survival of plants, as well as for maintaining oxygen levels in the Earth's
atmosphere.
```

3. Conversational Chain

This chain is used in conversational agents where maintaining context is critical. It leverages memory to store and recall previous interactions, enabling the model to respond in a way that reflects the ongoing conversation.

- **Usage**: Chatbots, virtual assistants, customer support applications, or any system requiring multiturn conversations
- **Components**: LLMs for generating responses, memory for storing context, and potentially external tools for more complex interactions

---

Note In the latest version of LangChain, you don't need to add the `openai_api_key` parameter anymore, but you need to define it as an environmental variable.

Example:

```
import os
# Set your OpenAI API key
os.environ["OPENAI_API_KEY"] = "sk-proj-056py5goMfqp8_g2gOgfhefr1HLriyWyP6
erQJ4dQyi3D2HWBxJgCWrjWMbvMTJdvxHlzaWm11T3BlbkFJss1mhhNZJ7YREWFugP2wKQoMHI
R3FMCDZxiOA_rPSrCfXZK6ZJbcGJ85dpMGV4adCt7R_zrUkA"
from langchain_core.output_parsers import StrOutputParser
from langchain_core.prompts import ChatPromptTemplate
from langchain_openai import ChatOpenAI
from langchain.memory import ConversationBufferMemory
from langchain.chains import LLMChain
# Step 1: Define a prompt template for conversation, using a variable for
user input
prompt = ChatPromptTemplate.from_messages(
    [("user", "{user_input}")]
)
# Step 2: Set up the ChatOpenAI model (gpt-3.5-turbo in this case) with
temperature control
llm = ChatOpenAI(model="gpt-3.5-turbo", temperature=0.7)
```

```
# Step 3: Create memory to store conversation history
memory = ConversationBufferMemory()

# Step 4: Create the chain combining prompt, model, and output parser

chain = LLMChain(prompt=prompt, llm=llm, memory=memory, output_
parser=StrOutputParser())
# Simulate a conversation by invoking the chain with memory
# First user input
response_1 = chain.invoke({"user_input": "Can you explain what
photosynthesis is?"})
print("AI Response 1:", response_1)
# Second user input
response_2 = chain.invoke({"user_input": "What happens during the light-
dependent reactions?"})
print("AI Response 2:", response_2)
# Third user input
response_3 = chain.invoke({"user_input": "Can you summarize both for me?"})
print("AI Response 3:", response_3)
print(memory)
```

Output:

```
AI Response 1: {'user_input': 'Can you explain what photosynthesis is?',
'history': '', 'text': 'Sure! Photosynthesis is the process by which green
plants, algae, and some bacteria convert light energy, usually from the
sun, into chemical energy in the form of glucose (sugar). This process
takes place in the chloroplasts of plant cells and involves the absorption
of carbon dioxide and water, which are converted into glucose and oxygen
through a series of complex chemical reactions. The glucose produced
through photosynthesis is used by the plant for energy and growth, while
the oxygen is released into the atmosphere as a byproduct. Photosynthesis
is essential for the survival of plants and other photosynthetic organisms,
as well as for the overall health of ecosystems.'}
AI Response 2: {'user_input': 'What happens during the light-dependent
reactions?', 'history': 'Human: Can you explain what photosynthesis is?\
nAI: Sure! Photosynthesis is the process by which green plants, algae, and
```

```
some bacteria convert light energy, usually from the sun, into chemical
energy in the form of glucose (sugar). This process takes place in the
chloroplasts of plant cells and involves the absorption of carbon dioxide
and water, which are converted into glucose and oxygen through a series of
complex chemical reactions. The glucose produced through photosynthesis is
used by the plant for energy and growth, while the oxygen is released into
the atmosphere as a byproduct. Photosynthesis is essential for the survival
of plants and other photosynthetic organisms, as well as for the overall
health of ecosystems.', 'text': 'During the light-dependent reactions,
also known as the light reactions, several key processes take place in the
thylakoid membranes of the chloroplast:\n\n1. Absorption of light: Light
energy is absorbed by chlorophyll and other pigments in the photosystems,
specifically Photosystem II and Photosystem I.\n\n2. Water splitting: The
absorbed light energy is used to split water molecules into oxygen, protons
(H+ ions), and electrons. This process releases oxygen as a byproduct.\n\
n3. Electron transport chain: The energized electrons from Photosystem II
are passed along a series of proteins in the electron transport chain,
generating ATP through the process of chemiosmosis.\n\n4. Production
of ATP and NADPH: The flow of electrons through the electron transport
chain ultimately leads to the production of ATP and NADPH, which are both
energy carriers used in the Calvin cycle.\n\nOverall, the light-dependent
reactions convert light energy into chemical energy in the form of ATP and
NADPH, which are then used in the Calvin cycle to produce glucose and other
organic compounds.'}
AI Response 3: {'user_input': 'Can you summarize both for me?', 'history':
'Human: Can you explain what photosynthesis is?\nAI: Sure! Photosynthesis
is the process by which green plants, algae, and some bacteria convert
light energy, usually from the sun, into chemical energy in the form of
glucose (sugar). This process takes place in the chloroplasts of plant
cells and involves the absorption of carbon dioxide and water, which are
converted into glucose and oxygen through a series of complex chemical
reactions. The glucose produced through photosynthesis is used by the plant
for energy and growth, while the oxygen is released into the atmosphere
as a byproduct. Photosynthesis is essential for the survival of plants
and other photosynthetic organisms, as well as for the overall health of
```

```
ecosystems.\nHuman: What happens during the light-dependent reactions?\
nAI: During the light-dependent reactions, also known as the light
reactions, several key processes take place in the thylakoid membranes
of the chloroplast:\n\n1. Absorption of light: Light energy is absorbed
by chlorophyll and other pigments in the photosystems, specifically
Photosystem II and Photosystem I.\n\n2. Water splitting: The absorbed light
energy is used to split water molecules into oxygen, protons (H+ ions),
and electrons. This process releases oxygen as a byproduct.\n\n3. Electron
transport chain: The energized electrons from Photosystem II are passed
along a series of proteins in the electron transport chain, generating ATP
through the process of chemiosmosis.\n\n4. Production of ATP and NADPH: The
flow of electrons through the electron transport chain ultimately leads
to the production of ATP and NADPH, which are both energy carriers used
in the Calvin cycle.\n\nOverall, the light-dependent reactions convert
light energy into chemical energy in the form of ATP and NADPH, which
are then used in the Calvin cycle to produce glucose and other organic
compounds.', 'text': 'Sure! The first passage discusses the importance of
self-care and setting boundaries to prevent burnout. It emphasizes the
need to prioritize mental and physical well-being in order to maintain a
healthy work-life balance.\n\nThe second passage highlights the benefits
of meditation for reducing stress and anxiety. It suggests incorporating
mindfulness practices into daily routines to improve overall mental health
and emotional well-being.'}
chat_memory=InMemoryChatMessageHistory(messages=[HumanMessage(conte
nt='Can you explain what photosynthesis is?', additional_kwargs={},
response_metadata={}), AIMessage(content='Sure! Photosynthesis is the
process by which green plants, algae, and some bacteria convert light
energy, usually from the sun, into chemical energy in the form of glucose
(sugar). This process takes place in the chloroplasts of plant cells
and involves the absorption of carbon dioxide and water, which are
converted into glucose and oxygen through a series of complex chemical
reactions. The glucose produced through photosynthesis is used by the
plant for energy and growth, while the oxygen is released into the
atmosphere as a byproduct. Photosynthesis is essential for the survival
of plants and other photosynthetic organisms, as well as for the overall
```

```
health of ecosystems.', additional_kwargs={}, response_metadata={}),
HumanMessage(content='What happens during the light-dependent reactions?',
additional_kwargs={}, response_metadata={}), AIMessage(content='During the
light-dependent reactions, also known as the light reactions, several key
processes take place in the thylakoid membranes of the chloroplast:\n\
n1. Absorption of light: Light energy is absorbed by chlorophyll and other
pigments in the photosystems, specifically Photosystem II and Photosystem
I.\n\n2. Water splitting: The absorbed light energy is used to split
water molecules into oxygen, protons (H+ ions), and electrons. This
process releases oxygen as a byproduct.\n\n3. Electron transport chain:
The energized electrons from Photosystem II are passed along a series
of proteins in the electron transport chain, generating ATP through the
process of chemiosmosis.\n\n4. Production of ATP and NADPH: The flow of
electrons through the electron transport chain ultimately leads to the
production of ATP and NADPH, which are both energy carriers used in the
Calvin cycle.\n\nOverall, the light-dependent reactions convert light
energy into chemical energy in the form of ATP and NADPH, which are then
used in the Calvin cycle to produce glucose and other organic compounds.',
additional_kwargs={}, response_metadata={}), HumanMessage(content='Can
you summarize both for me?', additional_kwargs={}, response_metadata={}),
AIMessage(content='Sure! The first passage discusses the importance of
self-care and setting boundaries to prevent burnout. It emphasizes the
need to prioritize mental and physical well-being in order to maintain a
healthy work-life balance.\n\nThe second passage highlights the benefits
of meditation for reducing stress and anxiety. It suggests incorporating
mindfulness practices into daily routines to improve overall mental health
and emotional well-being.', additional_kwargs={}, response_metadata={})])
```

4. Multi-input Chain

This type of chain accepts multiple inputs, which are processed either in parallel or in sequence depending on the workflow. It allows for more complex scenarios where various types of data or inputs must be handled together.

- **Usage**: When a task requires different sources of information, such as combining data from a user input and an external API or multiple models
- **Components**: Several input sources (e.g., a prompt and a knowledge base), multiple models, and tools to combine and process the inputs

Example:

```
from langchain_core.prompts import ChatPromptTemplate
from langchain_openai import ChatOpenAI
from langchain_core.output_parsers import StrOutputParser
from langchain.chains import SimpleSequentialChain

# Step 1: Define the first prompt to accept a question and context
question_prompt = ChatPromptTemplate.from_messages(
    [("user", "Given the context: '{context}', answer the question:
    '{question}'")]
)
# Step 2: Define the ChatOpenAI model
llm = ChatOpenAI(model="gpt-3.5-turbo", temperature=0.7)

# Step 3: Create the output parser
output_parser = StrOutputParser()

# Step 4: Combine the prompt and model into a chain
# This is a simple chain that handles multiple inputs (question and
context)
chain = question_prompt | llm | output_parser

# Step 5: Define the inputs for the multi-input chain
inputs = {
    "question": "How does photosynthesis work?",
    "context": "Photosynthesis is the process used by plants to convert
    light energy into chemical energy."
}
```

```
# Step 6: Run the chain with both inputs
response = chain.invoke(inputs)
# Output the response
print("Response:", response)
```

Output:

```
Response: Photosynthesis works by plants using sunlight to convert carbon
dioxide and water into glucose (sugar) and oxygen. This process takes place
in the chloroplasts of plant cells, where the green pigment chlorophyll
absorbs sunlight and initiates the chemical reactions that produce glucose.
The oxygen produced is released into the atmosphere as a byproduct.
```

Why Multi-input?

1. **Multi-input Prompt**: The `ChatPromptTemplate` defines a template that accepts two inputs: `context` and `question`. This prompt will insert both into the message for the language model.
2. **Model**: The `ChatOpenAI` model (`gpt-3.5-turbo`) is used to process the input and generate a response.
3. **Output Parser**: The `StrOutputParser` is used to parse the model's response into a string format. We will discuss the output parsers a bit later in the book.
4. **Chain Construction**: The chain combines the prompt, model, and output parser, handling both the question and context together as inputs.
5. **Invoke**: The `.invoke()` method is used to pass the inputs (both the question and the context) to the chain for processing.

5. Multi-output Chain

A multi-output chain takes an input and produces multiple outputs. This type of chain is useful when you want to generate different types of results based on a single input, such as extracting multiple pieces of information or generating multiple response options.

- **Usage**: Use cases where the same input must be processed in different ways, such as generating summaries, key takeaways, and action items from a single document
- **Components**: One input, multiple steps or LLM calls, and a set of outputs

```
from langchain_core.prompts import ChatPromptTemplate
from langchain_openai import ChatOpenAI
from langchain_core.output_parsers import StrOutputParser
from langchain.chains import LLMChain
from langchain.chains import SequentialChain

# Step 1: Define the prompt for generating a summary
summary_prompt = ChatPromptTemplate.from_messages(
    [("user", "Please summarize the following text: {input_text}")]
)

# Step 2: Define the prompt for extracting key points
key_points_prompt = ChatPromptTemplate.from_messages(
    [("user", "Extract the key points from the following text: {input_
    text}")]
)

# Step 3: Set up the ChatOpenAI model (same model for both tasks)
llm = ChatOpenAI(model="gpt-3.5-turbo", temperature=0.7)
# Step 4: Create the output parser
output_parser = StrOutputParser()

# Step 5: Create LLMChain for summarization and key point extraction
summary_chain = LLMChain(prompt=summary_prompt, llm=llm, output_
key="summary")  # Changed output key to "summary"
key_points_chain = LLMChain(prompt=key_points_prompt, llm=llm, output_
key="key_points")  # Changed output key to "key_points"

# Step 6: Create a SequentialChain that runs both chains (true
multi-output)
multi_output_chain = SequentialChain(
    chains=[summary_chain, key_points_chain],
```

```
    input_variables=["input_text"],  # single input passed to both chains
    output_variables=["summary", "key_points"]  # two outputs
)

# Step 7: Define the input text
input_text = """
Photosynthesis is a process used by plants to convert light energy into
chemical energy. During photosynthesis,
plants take in carbon dioxide (CO2) and water (H2O) from the air and soil.
Within the plant cell, the water is oxidized,
meaning it loses electrons, while the carbon dioxide is reduced, meaning it
gains electrons. This process converts
the water into oxygen and the carbon dioxide into glucose. The plant then
releases the oxygen back into the air,
and stores energy in the form of glucose molecules.
"""

# Step 8: Run the multi-output chain using apply() for multiple outputs
outputs = multi_output_chain.apply([{"input_text": input_text}])[0]

# Step 9: Output the responses
print("Summary:", outputs['summary'])
print("Key Points:", outputs['key_points'])
```

Output:

```
Summary: Photosynthesis is a process where plants convert light energy into
chemical energy by taking in carbon dioxide and water to produce oxygen and
glucose. The plant releases the oxygen and stores the glucose for energy.
Key Points: - Photosynthesis is a process used by plants to convert light
energy into chemical energy.
- Plants take in carbon dioxide and water from the air and soil.
- Water is oxidized and carbon dioxide is reduced during photosynthesis.
- The result is oxygen and glucose production.
- Oxygen is released back into the air, while glucose is stored as energy
  in the plant.
```

Why It's a Multi-output Chain

1. **Single Input**: The input (input_text) is passed once and processed through multiple chains.
2. **Multiple Outputs**: The input is processed in two different ways (summary and key points), and the outputs are stored in distinct keys (summary, key_points).
3. **Sequential Execution**: The SequentialChain ensures that both chains run in sequence, with the same input generating multiple outputs in a single invocation.

Handling Multiple Outputs with apply():

- Since SequentialChain supports multiple output variables, we use apply() instead of run() to handle cases where more than one output is generated. This is essential for returning a dictionary with multiple output keys.

6. Router Chain

The router chain acts as a decision-making hub that directs the input to different subchains based on predefined conditions or classifications. It's useful when you have various workflows that depend on the type of input.

- **Usage**: For tasks requiring conditional logic, such as routing customer queries to the right department (billing, technical support, etc.) or choosing the right model based on input complexity
- **Components**: A router module that decides which subchain to invoke, along with those subchains themselves

```
from langchain_core.output_parsers import StrOutputParser
from langchain_core.prompts import ChatPromptTemplate
from langchain_openai import ChatOpenAI

# Step 1: Define the prompts for summarization and key points extraction
```

```
# Summarization prompt
summary_prompt = ChatPromptTemplate.from_messages(
    [("user", "Please summarize the following text: {input_text}")]
)

# Key points extraction prompt
key_points_prompt = ChatPromptTemplate.from_messages(
    [("user", "Extract the key points from the following text: {input_
    text}")]
)
# Classifier prompt to determine if the input is asking for a "summary" or
"key points extraction"
classifier_prompt = ChatPromptTemplate.from_messages(
    [("user", "Classify this request as 'summarization' or 'key points
    extraction': {input_text}")]
)
# Step 2: Define the language model
llm = ChatOpenAI(model="gpt-3.5-turbo", temperature=0.7)

# Step 3: Define chains using the pipe operator
# Chain for classifying input
classifier_chain = classifier_prompt | llm | StrOutputParser()

# Chain for summarization
summary_chain = summary_prompt | llm | StrOutputParser()
# Chain for key points extraction
key_points_chain = key_points_prompt | llm | StrOutputParser()
# Step 4: Define a function to handle the routing based on classification
def router_chain(input_text):
    # Classify the input (is it a request for summarization or key points?)
    classification = classifier_chain.invoke({"input_text": input_text})
    # Route to the appropriate chain based on the classification result
    if "summarization" in classification.lower():
        return summary_chain.invoke({"input_text": input_text})
    elif "key points extraction" in classification.lower():
        return key_points_chain.invoke({"input_text": input_text})
```

```
    else:
        # Fallback to the summary chain if the classification is unclear
        return summary_chain.invoke({"input_text": input_text})

# Step 5: Define input texts
input_text_1 = "Summarize this text: Photosynthesis is a process used by
plants to convert light energy into chemical energy."
input_text_2 = "Give me the key points of the following text:
Photosynthesis is a process used by plants to convert light energy into
chemical energy."

# Step 6: Run the router chain on different inputs
output_1 = router_chain(input_text_1)
output_2 = router_chain(input_text_2)

# Step 7: Print the outputs
print("Output 1:", output_1)
print("Output 2:", output_2)
```

Output:

```
Output 1: Photosynthesis is the process that plants use to convert light
energy into chemical energy.
Output 2: - Photosynthesis is a process used by plants
- Plants convert light energy into chemical energy through photosynthesis
```

Why Router Chain?

- **Prompt Definition:** Each prompt is defined using ChatPromptTemplate.from_messages(). This includes the summary_prompt, key_points_prompt, and classifier_prompt for routing.
- **Chained Operations:** The chains (classifier_chain, summary_chain, and key_points_chain) are created using the pipe (|) operator to chain together the prompt, model (ChatOpenAI), and output parser (StrOutputParser).
- **Router Function:** The router_chain function first invokes the classifier_chain to classify the input as either a “summarization” or “key points extraction” task.

- Based on the classification result, it dynamically routes the input to the appropriate chain (summary_chain or key_points_chain). If the classification is unclear, it defaults to the summarization chain.
- **Running the Chain:** The router_chain function is run on two different inputs, input_text_1 and input_text_2, and the outputs are printed.

7. Control Flow Chain

A control flow chain allows branching and conditional execution based on the results of intermediate steps. The workflow can change dynamically depending on the decisions made at each stage, enabling complex reasoning processes.

- **Usage**: Scenarios where certain actions are taken only if specific conditions are met, such as checking the confidence level of a model's output or validating an API response
- **Components**: Logic that governs branching (e.g., if-else statements), conditional steps, and error handling mechanisms

```
from langchain_core.output_parsers import StrOutputParser
from langchain_core.prompts import ChatPromptTemplate
from langchain_openai import ChatOpenAI
# Step 1: Define prompts for different tasks
# Prompt to answer a definition-related question
definition_prompt = ChatPromptTemplate.from_messages(
    [("user", "Define the following concept: {concept}")]
)

# Prompt to perform a calculation
calculation_prompt = ChatPromptTemplate.from_messages(
    [("user", "Calculate the following: {calculation}")]
)

# Classifier prompt to determine if the input is asking for a "definition"
or a "calculation"
classifier_prompt = ChatPromptTemplate.from_messages(
```

```
    [("user", "Classify this request as 'definition' or 'calculation':
    {input_text}")]
)

# Step 2: Set up the ChatOpenAI model
llm = ChatOpenAI(model="gpt-3.5-turbo", temperature=0.7)
# Step 3: Define chains using the pipe operator

# Chain for classifying input
classifier_chain = classifier_prompt | llm | StrOutputParser()

# Chain for definition tasks
definition_chain = definition_prompt | llm | StrOutputParser()

# Chain for calculation tasks
calculation_chain = calculation_prompt | llm | StrOutputParser()

# Step 4: Define a function to handle control flow based on classification
def control_flow_chain(input_text):
    # Classify the input (is it a request for a definition or a
      calculation?)
    classification = classifier_chain.invoke({"input_text": input_text})
    # Route to the appropriate chain based on the classification result
    if "definition" in classification.lower():
        concept = input_text.replace("Define", "").strip()
        return definition_chain.invoke({"concept": concept})
    elif "calculation" in classification.lower():
        calculation = input_text.replace("Calculate", "").strip()
        return calculation_chain.invoke({"calculation": calculation})
    else:
        # Default response if classification is unclear
        return "Sorry, I didn't understand your request."

# Step 5: Define input texts
input_text_1 = "Define photosynthesis"
input_text_2 = "Calculate 5 + 3"
```

```
# Step 6: Run the control flow chain on different inputs
output_1 = control_flow_chain(input_text_1)
output_2 = control_flow_chain(input_text_2)

# Step 7: Print the outputs
print("Output 1:", output_1)
print("Output 2:", output_2)
```

Output:

```
Output 1: Photosynthesis is the process by which green plants, algae, and
some bacteria convert light energy, usually from the sun, into chemical
energy in the form of glucose. This process involves the absorption of
carbon dioxide and water, which are then converted into oxygen and glucose
through a series of chemical reactions. Oxygen is released as a byproduct
of this process, making photosynthesis essential for the survival of most
living organisms on Earth.
Output 2: 5 + 3 = 8
```

Key Features of the Control Flow Chain

1. **Conditional Logic**: The input is processed using conditional logic to determine which chain (definition or calculation) should handle the request.
2. **Dynamic Routing**: Based on the classification result, the input is dynamically routed to the appropriate chain.
3. **Flexible Task Handling**: This control flow chain can easily be extended to handle more types of inputs, making it a versatile way to manage tasks based on user requests.

8. Retrieval-Aware Chain

This chain is integrated with a retrieval mechanism, such as a vector database or a search engine, to retrieve relevant information before making decisions or generating responses. It's typically used in situations where context or additional data is needed to complete the task.

- **Usage**: Question-answering systems that need to pull information from knowledge bases or document repositories to provide accurate answers
- **Components**: A retrieval component (e.g., vector search or document retrieval) combined with LLM calls to process the retrieved information

---

Note For the next example, you need to run the command `pip install faiss-gpu` as we use faiss.

Example:

```
from langchain_core.output_parsers import StrOutputParser
from langchain_core.prompts import ChatPromptTemplate
from langchain_openai import ChatOpenAI
from langchain.vectorstores import FAISS
from langchain.embeddings.openai import OpenAIEmbeddings
# Step 1: Set up the FAISS vector store with embeddings
# This example assumes the OpenAI API is configured and available
# Define some documents (texts) related to quantum computing
documents = [
    "Quantum computing is a type of computation that harnesses the
    collective properties of quantum states.",
    "Quantum computers use quantum bits, or qubits, which can represent and
    store more information than classical bits.",
    "The fundamental principle of quantum computing is superposition, which
    allows qubits to be in multiple states at once.",
    "Entanglement is another key property of quantum computing, allowing
    qubits to be interconnected no matter the distance."
]

# Step 2: Embed the documents using OpenAI embeddings
embeddings = OpenAIEmbeddings()  # Ensure you have OpenAI API keys
                                   configured
```

```
# Step 3: Create a FAISS vector store from the documents and their
embeddings
vector_store = FAISS.from_texts(documents, embeddings)

# Step 4: Define the prompt that will use the retrieved context
retrieval_prompt = ChatPromptTemplate.from_messages(
    [("user", "Given the following context, answer the question:
    {context}")]
)

# Step 5: Define the ChatOpenAI model
llm = ChatOpenAI(model="gpt-3.5-turbo", temperature=0.7)
# Step 6: Define the retrieval-aware chain using FAISS
def retrieval_aware_chain(input_query):
    # Step 6.1: Retrieve relevant documents based on the query
    retrieved_documents = vector_store.similarity_search(input_query)
    # FAISS similarity search
    context = " ".join([doc.page_content for doc in retrieved_
    documents])  # Combine documents into a single context
    # Step 6.2: Run the LLM chain with the retrieved context
    response = (retrieval_prompt | llm | StrOutputParser()).
    invoke({"context": context})
    return response

# Step 7: Define an input query
input_query = "What is quantum entanglement?"

# Step 8: Run the retrieval-aware chain
output = retrieval_aware_chain(input_query)

# Step 9: Print the output
print("Output:", output)
```

Output:

```
Output: What is entanglement in quantum computing?
Entanglement is a key property of quantum computing that allows qubits to
be interconnected no matter the distance.
```

9. Agent Chain

An agent chain is designed to allow a language model to interact with multiple tools or APIs autonomously. The LLM acts as an agent, deciding which tool to use and when, allowing for highly dynamic workflows where the model selects the appropriate actions.

- **Usage**: Complex applications where the model must autonomously decide which action to take, such as querying an API, searching a database, or executing a code snippet
- **Components**: The agent (LLM) interacts with external tools, APIs, or modules and follows predefined logic or dynamically generated plans

Example:

```
from langchain_core.output_parsers import StrOutputParser
from langchain_core.prompts import ChatPromptTemplate
from langchain_openai import ChatOpenAI
from langchain.tools import Tool, BaseTool
from langchain.vectorstores import FAISS
from langchain.embeddings.openai import OpenAIEmbeddings
import math

# Step 1: Define the tools (calculator and retrieval tool)

# Define a calculator tool to perform basic math operations
class CalculatorTool(BaseTool):
    def _run(self, input_query: str) -> str:
        """Run the calculator tool to perform basic arithmetic."""
        try:
            # Extract the mathematical expression by removing "calculate"
              or "Calculate"
            expression = input_query.lower().replace("calculate", "").
            strip()
            return str(eval(expression))  # Use eval safely for basic
            calculations
        except Exception:
            return "Invalid calculation."
```

```
    def name(self):
        return "calculator"

    def description(self):
        return "A simple calculator tool for performing basic arithmetic
        operations."

# Create an instance of CalculatorTool
calculator_tool = CalculatorTool()

# Define the FAISS-based retrieval tool for information retrieval
documents = [
    "Quantum computing is a type of computation that harnesses the
    collective properties of quantum states.",
    "Quantum computers use quantum bits, or qubits, which can represent and
    store more information than classical bits.",
    "The fundamental principle of quantum computing is superposition, which
    allows qubits to be in multiple states at once.",
    "Entanglement is another key property of quantum computing, allowing
    qubits to be interconnected no matter the distance."
]

embeddings = OpenAIEmbeddings()
vector_store = FAISS.from_texts(documents, embeddings)

class RetrievalTool(BaseTool):
    def _run(self, input_query: str) -> str:
        """Run the retrieval tool to search the vector store for relevant
        information."""
        retrieved_documents = vector_store.similarity_search(input_query)
        return " ".join([doc.page_content for doc in retrieved_documents])

    def name(self):
        return "retrieval"

    def description(self):
        return "A tool for retrieving information about quantum computing."
```

```
# Create an instance of RetrievalTool
retrieval_tool = RetrievalTool()

# Step 2: Define the agent prompt with explicit instructions
agent_prompt = ChatPromptTemplate.from_messages(
    [("user", "If the query asks to perform a calculation (e.g., 'calculate
    5 + 7'), respond with 'calculate'. "
              "If the query asks for information (e.g., 'What is quantum
              computing?'), respond with 'retrieve'. "
              "Input: {input_query}")]
)

# Step 3: Define the ChatOpenAI model (the agent)
llm = ChatOpenAI(model="gpt-3.5-turbo", temperature=0.7)

# Step 4: Define the agent chain function
def agent_chain(input_query):
    # Ask the agent to classify the task (calculation or retrieval)
    agent_decision = (agent_prompt | llm | StrOutputParser()).
    invoke({"input_query": input_query})

    # Based on the agent's decision, invoke the appropriate tool
    if "calculate" in agent_decision.lower():
        return calculator_tool._run(input_query)
    else:
        return retrieval_tool._run(input_query)

# Step 5: Define the input queries
input_query_1 = "Calculate 5 + 7"
input_query_2 = "Explain quantum superposition"

# Step 6: Run the agent chain on different inputs
output_1 = agent_chain(input_query_1)  # Expecting a calculation result
output_2 = agent_chain(input_query_2)  # Expecting information
retrieval result

# Step 7: Print the outputs
print("Output 1:", output_1)
print("Output 2:", output_2)
```

Output:

```
Output 1: 12
Output 2: The fundamental principle of quantum computing is superposition,
which allows qubits to be in multiple states at once. Quantum computing is
a type of computation that harnesses the collective properties of quantum
states. Entanglement is another key property of quantum computing, allowing
qubits to be interconnected no matter the distance. Quantum computers use
quantum bits, or qubits, which can represent and store more information
than classical bits.
```

Breakdown of Key Steps of This More Complicated Code

- **CalculatorTool Definition:** A class CalculatorTool is defined, inheriting from BaseTool.
- **The _run() method is implemented, which**
    - Strips the word "calculate" from the input query.
    - Evaluates the remaining mathematical expression (e.g., "5 + 7") using eval().
    - Returns the result of the calculation.
    - If the evaluation fails (e.g., for invalid expressions), it returns "Invalid calculation."
- **RetrievalTool Definition**
    - A class RetrievalTool is defined, inheriting from BaseTool.
    - The _run() method is implemented, which
        - Uses FAISS to perform a similarity search based on the input query (e.g., "Explain quantum superposition").
        - Retrieves relevant documents from the vector store.
        - Concatenates the content of the retrieved documents into a single response.

- **Embedding and Vector Store Setup**
    - A list of documents related to quantum computing is created.
    - OpenAIEmbeddings are used to embed these documents into vectors.
    - The document embeddings are stored in a FAISS vector store, which allows for similarity-based document retrieval.
- **Agent Prompt Setup**
    - **The agent prompt is defined, providing explicit instructions to the language model:**
        - If the input asks for a calculation (e.g., "Calculate 5 + 7"), the model should respond with "calculate."
        - If the input asks for information (e.g., "What is quantum superposition?"), the model should respond with "retrieve."
- **Agent Chain Function**
    - **The function agent_chain(input_query) performs the following steps:**
        - Passes the input query to the agent prompt (via the language model).
        - The agent responds with either "calculate" or "retrieve," based on the task.
    - **Depending on the agent's decision:**
        - If "calculate" is returned, it calls the CalculatorTool to perform the calculation.
        - If "retrieve" is returned, it calls the RetrievalTool to fetch relevant information from the FAISS vector store.

10. Parallel Chain

A parallel chain allows multiple processes to run simultaneously, with their results combined at the end. This can improve efficiency when independent tasks can be processed at the same time.

- **Usage**: Situations where different tasks or models can be executed in parallel, such as generating multiple drafts of a text or performing several independent API calls
- **Components**: Multiple parallel operations that feed into a final aggregation or decision step

Example:

```
from langchain_core.prompts import ChatPromptTemplate
from langchain_core.runnables import RunnableParallel
from langchain_openai import ChatOpenAI

# Step 1: Set up the OpenAI model
model = ChatOpenAI()

# Step 2: Define the chains for independent tasks

# Chain to summarize a concept
summarize_chain = ChatPromptTemplate.from_template("Summarize the concept
of {concept}") | model

# Chain to provide detailed information about the concept
information_chain = ChatPromptTemplate.from_template("Provide detailed
information about {concept}") | model

# Step 3: Set up the parallel chain to run both tasks concurrently
parallel_chain = RunnableParallel(summary=summarize_chain,
information=information_chain)

# Step 4: Define the input concept
input_concept = {"concept": "Quantum computing"}

# Step 5: Run the parallel chain with the input concept
outputs = parallel_chain.invoke(input_concept)
```

```
# Step 6: Print the outputs
print("Summary Output:", outputs["summary"])
print("Information Output:", outputs["information"])
```

Output:

```
Summary Output: content='Quantum computing is a type of computing that
utilizes the principles of quantum mechanics to perform operations on
data. Unlike classical computing, which uses bits as the fundamental unit
of information, quantum computing uses quantum bits, or qubits, which can
exist in multiple states simultaneously. This allows quantum computers to
perform complex calculations much faster than classical computers, making
them potentially capable of solving problems that are currently infeasible
with traditional computing methods.' additional_kwargs={'refusal': None}
response_metadata={'token_usage': {'completion_tokens': 85, 'prompt_
tokens': 15, 'total_tokens': 100, 'completion_tokens_details': {'audio_
tokens': None, 'reasoning_tokens': 0}, 'prompt_tokens_details': {'audio_
tokens': None, 'cached_tokens': 0}}, 'model_name': 'gpt-3.5-turbo-0125',
'system_fingerprint': None, 'finish_reason': 'stop', 'logprobs': None}
id='run-748514c4-35eb-4ccc-a3a8-68dee2c3fa74-0' usage_metadata={'input_
tokens': 15, 'output_tokens': 85, 'total_tokens': 100, 'input_token_
details': {'cache_read': 0}, 'output_token_details': {'reasoning': 0}}
Information Output: content='Quantum computing is a type of computing
that uses quantum-mechanical phenomena, such as superposition and
entanglement, to perform operations on data. Unlike classical computing,
which uses bits to represent data as either 0 or 1, quantum computing
uses quantum bits, or qubits, which can exist in multiple states
simultaneously due to superposition.\n\nOne of the key principles of
quantum computing is superposition, which allows qubits to exist in
a state that is a combination of both 0 and 1 at the same time. This
enables quantum computers to perform multiple calculations simultaneously,
making them potentially much faster than classical computers for certain
types of problems.\n\nAnother important concept in quantum computing is
entanglement, which allows qubits to be correlated with each other in
such a way that the state of one qubit can instantly affect the state of
another, regardless of the distance between them. This property enables
```

```
quantum computers to perform certain types of operations more efficiently
than classical computers.\n\nQuantum computing has the potential to
revolutionize fields such as cryptography, drug discovery, optimization,
and machine learning by solving complex problems that are currently
infeasible for classical computers. However, quantum computers are
still in the early stages of development and face significant technical
challenges, such as maintaining the coherence of qubits and scaling up
to larger systems.\n\nCompanies such as IBM, Google, and Microsoft are
investing heavily in quantum computing research and development, and there
are also startups and research institutions around the world working on
advancing the field. As quantum computing continues to progress, it holds
the promise of enabling breakthroughs in a wide range of scientific and
technological applications.' additional_kwargs={'refusal': None} response_
metadata={'token_usage': {'completion_tokens': 326, 'prompt_tokens':
13, 'total_tokens': 339, 'completion_tokens_details': {'audio_tokens':
None, 'reasoning_tokens': 0}, 'prompt_tokens_details': {'audio_tokens':
None, 'cached_tokens': 0}}, 'model_name': 'gpt-3.5-turbo-0125', 'system_
fingerprint': None, 'finish_reason': 'stop', 'logprobs': None} id='run-
e5913f67-0055-4117-8d86-a5ba913e2dc3-0' usage_metadata={'input_tokens': 13,
'output_tokens': 326, 'total_tokens': 339, 'input_token_details': {'cache_
read': 0}, 'output_token_details': {'reasoning': 0}}
```

What Does the Code Do?

1. **Model Setup**
    - ChatOpenAI() is instantiated to serve as the language model for both tasks (summarization and detailed information retrieval).
2. **Chain Definitions**
    - **summarize_chain**: A prompt asks the model to summarize the given concept (e.g., “Quantum computing”).
    - **information_chain**: A prompt asks the model to provide detailed information about the same concept.

3. **Parallel Execution with RunnableParallel**
    - **RunnableParallel** is used to execute both chains concurrently.
    - Two chains are passed as arguments (summary for the summarization chain and information for the detailed information chain), which will run in parallel.
4. **Input Concept**
    - The input concept is a dictionary containing the key "concept" with the value "Quantum computing."
    - This input is passed to both chains.
5. **Running the Chains in Parallel**
    - The invoke() method is called on parallel_chain to execute both chains concurrently.
    - The outputs are returned as a dictionary with keys "summary" and "information."
6. **Outputs**
    - The outputs from both chains (summary and detailed information) are printed.

Key Features of RunnableParallel

- **Concurrent Execution:** Both chains are executed concurrently, reducing the overall time required for execution.
- **Flexible Input Handling:** The same input ("Quantum computing") is passed to both chains, but you can modify it to handle different inputs for each chain if needed.
- **Combined Outputs:** The results from both chains are combined into a single output dictionary, where each key corresponds to the respective chain's output.

11. Custom Chain

Custom chains are tailored to the specific needs of an application, combining various components in novel ways. Developers can create a custom sequence of operations that fit their unique use case, combining steps from different chain types into a bespoke workflow.

- **Usage**: When none of the prebuilt chain types meet the specific requirements of the task, and custom logic, steps, or external integrations are needed
- **Components**: A combination of modules, tools, logic, and LLMs to suit the custom requirements of the application

These **LangChain chain types** provide a flexible framework for building diverse and sophisticated workflows tailored to the specific needs of different applications. By combining or modifying these chain types, developers can orchestrate complex interactions and achieve nuanced, multistep tasks when working with large language models.

Example:

```
from langchain_core.prompts import ChatPromptTemplate
from langchain_core.runnables import Runnable
from langchain_openai import ChatOpenAI

# Step 1: Set up the OpenAI model
model = ChatOpenAI()

# Step 2: Define the chain for summarizing the concept
summarize_chain = ChatPromptTemplate.from_template("Summarize the concept
of {concept}") | model

# Step 3: Define the chain for generating a quiz question based on
the summary
quiz_chain = ChatPromptTemplate.from_template("Create a quiz question based
on the summary: {summary}") | model

# Step 4: Create a custom chain that first summarizes, then
generates a quiz
class CustomChain(Runnable):
```

```
    def invoke(self, input_data):
        # First, get the summary of the concept
        summary = summarize_chain.invoke({"concept": input_
        data["concept"]})

        # Then, use the summary to generate a quiz question
        quiz_question = quiz_chain.invoke({"summary": summary})

        # Return both the summary and the quiz question
        return {"summary": summary, "quiz_question": quiz_question}

# Step 5: Create an instance of the custom chain
custom_chain = CustomChain()

# Step 6: Define the input concept
input_concept = {"concept": "Quantum computing"}

# Step 7: Run the custom chain with the input concept
output = custom_chain.invoke(input_concept)

# Step 8: Print the outputs
print("Summary Output:", output["summary"])
print("Quiz Question Output:", output["quiz_question"])
```

Output:

```
Summary Output: content='Quantum computing is a type of computing that
uses quantum-mechanical phenomena, such as superposition and entanglement,
to perform operations on data. This allows quantum computers to process
information much faster than classical computers. Quantum computing has the
potential to revolutionize fields such as cryptography, optimization, and
drug discovery by solving complex problems that are currently intractable
for classical computers.' additional_kwargs={'refusal': None} response_
metadata={'token_usage': {'completion_tokens': 76, 'prompt_tokens':
15, 'total_tokens': 91, 'completion_tokens_details': {'audio_tokens':
None, 'reasoning_tokens': 0}, 'prompt_tokens_details': {'audio_tokens':
None, 'cached_tokens': 0}}, 'model_name': 'gpt-3.5-turbo-0125', 'system_
fingerprint': None, 'finish_reason': 'stop', 'logprobs': None} id='run-7
bc1450c-9826-4b69-8677-7d76f6cba1f7-0' usage_metadata={'input_tokens': 15,
```

```
'output_tokens': 76, 'total_tokens': 91, 'input_token_details': {'cache_
read': 0}, 'output_token_details': {'reasoning': 0}}
Quiz Question Output: content='How does quantum computing utilize
superposition and entanglement to perform operations on data, and
what advantages does this offer over classical computing methods?'
additional_kwargs={'refusal': None} response_metadata={'token_usage':
{'completion_tokens': 28, 'prompt_tokens': 284, 'total_tokens': 312,
'completion_tokens_details': {'audio_tokens': None, 'reasoning_tokens':
0}, 'prompt_tokens_details': {'audio_tokens': None, 'cached_tokens':
0}}, 'model_name': 'gpt-3.5-turbo-0125', 'system_fingerprint': None,
'finish_reason': 'stop', 'logprobs': None} id='run-9cf17c37-ac7c-48ab-b468-
c6ef3b7389f5-0' usage_metadata={'input_tokens': 284, 'output_tokens': 28,
'total_tokens': 312, 'input_token_details': {'cache_read': 0}, 'output_
token_details': {'reasoning': 0}}
```

Key Features of a Custom Chain

- **Custom Processing Logic:** The CustomChain class defines a two-step process: first generating a summary and then creating a quiz question based on the summary.
- **Sequential Execution:** The chain runs each step in sequence, passing the result of one step (summary) into the next step (quiz question generation).
- **Combined Outputs:** The chain returns both outputs (summary and quiz question) in a single response.

What Does This Code Do?

- **Model Setup:** Initializes a ChatOpenAI model to handle both summarization and quiz generation tasks
- **Summarization Chain:** Defines a chain (summarize_chain) that generates a summary of a concept based on a given input (e.g., "Quantum computing")
- **Quiz Generation Chain:** Defines a chain (quiz_chain) that creates a quiz question based on the summary of the concept

- **CustomChain Class**
    - **Step 1:** Generates a summary of the concept using the summarize_chain
    - **Step 2:** Uses the summary to generate a quiz question with the quiz_chain
    - **Combined Output:** Returns both the summary and the quiz question as output
- **Execution:** Runs the custom chain by passing the concept ("Quantum computing"), and the chain outputs both a summary and a quiz question
- **Outputs:** Prints the generated summary and the quiz question based on the input concept

Conclusion

In this chapter, we covered the basics of LangChain and its integration with Python for building advanced NLP applications. We explored key components such as chains, prompts, memory, and tools, which enable developers to create flexible and scalable workflows. LangChain simplifies the process of working with large language models, allowing for efficient management of context and multistep processing.

By mastering these fundamental concepts, you are now equipped to build a variety of language model-based applications, from simple chatbots to more complex data retrieval systems.

In the next chapter, we'll dive deeper into more advanced components and conceptions like **LangChain Memory**, which enables models to retain information across interactions. We'll also explore **agents and tools** in LangChain, which allow dynamic decision-making, and discuss **indexes and retrievers**, essential for handling large datasets efficiently. These advanced features will help you build even more powerful and context-aware NLP applications.

CHAPTER 2

LangChain and Python: Advanced Components

As the fields of machine learning and natural language processing continue to advance, Python remains at the heart of innovation, providing a robust ecosystem of tools, libraries, and frameworks. Among these, LangChain has emerged as a powerful framework tailored specifically to streamline and enhance workflows around large language models (LLMs). While foundational components of LangChain simplify common tasks such as chaining models, querying, and prompt management, there exists an extensive suite of advanced components that significantly expands LangChain's utility. This chapter delves into these advanced features, guiding readers through their purpose, application, and implementation in Python to tackle complex LLM workflows effectively.

LangChain's advanced components, including tools for memory management, custom agent creation, tools, indexes, and retrievers, provide practitioners with a sophisticated toolkit that caters to varied and challenging use cases. These components allow developers to push beyond basic model interactions, enabling functionalities such as real-time memory recall, multiagent systems, and seamless integration of external data sources, each enhancing the adaptability and intelligence of LLM-based applications.

In this chapter, we will explore these advanced components in-depth, breaking down their architecture, discussing best practices, and showcasing practical applications with Python. By the end of this chapter, readers will be equipped with the knowledge to leverage LangChain's full potential in developing customized, resilient, and intelligent language model applications.

D. Grigorov, *Intermediate Python and Large Language Models*, https://doi.org/10.1007/979-8-8688-1475-4_2

We begin by highlighting the evolving role of Python in AI and introducing LangChain as a powerful framework for building sophisticated LLM-based applications. The introduction sets the stage for exploring the advanced tools and capabilities that LangChain offers.

- **Python's Role in AI and NLP**

  Python remains the foundational language driving innovation in machine learning and natural language processing.

- **Introduction to LangChain**

  LangChain is presented as a framework designed to streamline the development of applications powered by large language models (LLMs).

- **Beyond the Basics**

  While LangChain simplifies core tasks like chaining and prompt management, this chapter focuses on its **advanced components**, including

  - Memory systems
  - Custom agents
  - External tools
  - Indexes and retrievers

- **Capabilities of Advanced Components**

  These tools enable

  - Real-time memory recall
  - Multiagent systems
  - Contextual and personalized interactions
  - Integration with diverse external data sources

LangChain Memory

In developing applications that harness large language models (LLMs), a common challenge is enabling these models to "remember" past interactions, mimicking conversational context and continuity. **LangChain Memory addresses** this by providing mechanisms to store, retrieve, and utilize conversation history within LangChain workflows. Unlike traditional stateless models, memory-enabled systems can reference past exchanges, allowing them to maintain a consistent narrative, track user preferences, and dynamically adapt responses over time.

This subtopic covers LangChain's memory capabilities, exploring different memory types (short-term, long-term, and specialized memory modules) and demonstrating how each can enhance interactive applications. From personalizing user interactions to facilitating complex dialogues in customer service or education, LangChain Memory is a transformative tool for developing applications that feel more intuitive and responsive to users.

Understanding LangChain's Memory Module

In LangChain, the Memory module plays a foundational role in enabling large language models (LLMs) to retain information between calls of a chain or agent. This persistence of state is essential in scenarios where the language model benefits from remembering past interactions, allowing it to make more contextually relevant and informed decisions.

By offering a standard interface for storing and retrieving information across interactions, LangChain's Memory module allows developers to equip language models with memory and continuity. This ability to remember is invaluable for applications such as personal assistants, autonomous agents, and agent-based simulations, where the model needs to retain user preferences, previous queries, or other critical details over time.

Key Capabilities of the Memory Module

The Memory module enables an LLM to maintain a running context by storing user inputs, system responses, and any other relevant information from past interactions. This stored data can then be accessed in future interactions, giving the model a sense of continuity and memory, which results in more accurate, contextually aware responses and decisions.

Why Memory Matters

The Memory module transforms a language model from a reactive agent into one that can adapt and respond based on past interactions. This continuity is crucial for creating interactive and personalized applications. With memory, the language model can provide richer responses by leveraging prior knowledge, which is particularly valuable in applications like personal assistants, customer support agents, and educational tutors.

When to Use the Memory Module

Use the Memory module whenever you want to build applications requiring context and continuity across interactions. For instance, a personal assistant application would benefit from memory as it allows the model to retain user preferences, recall previous questions, and track ongoing issues. Similarly, in autonomous agents and simulations, memory allows the model to make decisions that reflect accumulated knowledge, making interactions feel more coherent and informed.

Core Processes in the Memory System: Reading and Writing

Each memory system within LangChain performs two essential tasks: reading from memory and writing to memory. During any run, the model accesses its memory system at two key points:

- **Reading from Memory:** Before executing its main logic, the model reads stored information to augment user inputs, allowing it to make more informed decisions during processing.
- **Writing to Memory:** After generating a response, the model records the details of the current interaction to memory, ensuring that this information is available for future reference.

These read and write operations make it possible for the model to maintain context across interactions, giving it the ability to build on prior knowledge.

Structuring a Memory System

When designing a memory system, two core considerations come into play:

- **State Storage Method:** At the heart of the memory system is a record of all chat interactions. LangChain's memory module provides flexibility in how these interactions are stored, ranging from temporary in-memory lists for quick access to persistent database solutions for long-term storage.
- **State Querying Approach:** Storing chat logs is straightforward; the challenge lies in developing algorithms to interpret these logs meaningfully. A basic memory system might simply display recent messages, while a more sophisticated system might summarize the last "K" interactions. The most advanced systems can even identify entities from stored chats and retrieve relevant details about those entities when needed in the current session. This adaptability allows developers to tailor the memory query method to the specific needs of the application.

LangChain's Memory module offers a straightforward setup for initiating basic memory systems while supporting the creation of more advanced and customized systems as necessary.

By incorporating LangChain's Memory module, developers can create language model-driven applications that are not only responsive and adaptive but also capable of continuous learning and refinement. This module equips LLMs with memory and context, making them more capable, personalized, and effective in delivering consistent, user-centric experiences.

Note LangChain Memory is a powerful feature designed initially to enhance chatbots' functionality, by enabling them to retain context and significantly improve their conversational capabilities. Traditionally, chatbots process each user prompt independently, without considering the history of interactions. This isolated approach often results in responses that lack continuity, leading to disjointed and sometimes unsatisfying user experiences. LangChain addresses this challenge by offering dedicated memory components that manage and utilize previous chat messages, seamlessly integrating them into conversational chains. This functionality is vital for creating chatbots that need to remember prior interactions, allowing them to provide coherent and contextually relevant responses that feel more natural and engaging to users.

LangChain Memory Types

LangChain offers a rich suite of memory types that equip language models with the ability to remember, recall, and integrate contextual information from prior interactions. Each memory type is uniquely suited for different use cases, ranging from simple chat histories to complex knowledge-based and entity-driven contexts. These options allow developers to build applications with varying levels of depth, persistence, and relational awareness, creating personalized, coherent, and dynamic user experiences.

Here's an in-depth look at each type of memory offered by LangChain.

ConversationBufferMemory

ConversationBufferMemory is a straightforward memory type that stores a verbatim transcript of all interactions within a session. This approach maintains a full conversation history, allowing the language model to reference any part of the ongoing exchange and to provide contextually aware responses.

- **Use Case:** Applications where a complete record of interactions is valuable, such as detailed customer support systems, coaching applications, and collaborative brainstorming tools.
- **Advantages:** By keeping all interactions in memory, the model can access comprehensive context, which helps ensure consistent responses.
- **Limitations:** For long or continuous interactions, storing a full transcript can become resource-intensive, potentially leading to performance issues if not managed correctly. One drawback is that it retains the complete interaction history (up to the maximum token limit supported by the specific LLM), which means that for each new

question, the entire prior discussion is sent to the LLM API as tokens. This can lead to significant costs, as API usage fees are based on the total number of tokens processed per interaction. Additionally, as the conversation grows, this can introduce latency, impacting the model's response time due to the increasing amount of data being processed with each API call.

Example:

```
from langchain.chat_models import ChatOpenAI
from langchain.prompts import ChatPromptTemplate,
SystemMessagePromptTemplate, HumanMessagePromptTemplate
from langchain.chains import LLMChain
from langchain.memory import ConversationBufferMemory

# Initialize the chat model
chat_model = ChatOpenAI(model="gpt-3.5-turbo", temperature=0.7)

# Define the prompt templates
system_prompt = SystemMessagePromptTemplate.from_template("You are a
helpful assistant.")
human_prompt = HumanMessagePromptTemplate.from_template("{history}\n\nUser:
{input}")

# Wrap prompts in a ChatPromptTemplate
chat_prompt = ChatPromptTemplate.from_messages([system_prompt, human_
prompt])

# Set up the memory
memory = ConversationBufferMemory(return_messages=True)

# Create the chain with memory
conversation_chain = LLMChain(
    llm=chat_model,
    prompt=chat_prompt,
    memory=memory
)
# Example interaction 1
user_input_1 = "Hello, can you help me with some Python code?"
```

```
response_1 = conversation_chain.run(input=user_input_1)
print(response_1)
# Example interaction 2
user_input_2 = "I need help with writing a loop."
response_2 = conversation_chain.run(input=user_input_2)
print(response_2)
# Example interaction 3
user_input_3 = "Thanks! How do I make it run faster?"
response_3 = conversation_chain.run(input=user_input_3)
print(response_3)
```

Output:

```
Of course! I'd be happy to help. What do you need assistance with
in Python?
Of course! What kind of loop are you trying to write in Python? Do you have
a specific task or problem that you need help with? Let me know the details
so I can assist you better.
There are several ways you can optimize your Python code to make it run
faster. Here are some tips:
1. **Use appropriate data structures**: Choose the right data structure for
your task. For example, if you need to perform a lot of lookups, consider
using a dictionary instead of a list.
2. **Avoid unnecessary operations**: Make sure your code is not performing
redundant calculations or operations that can be eliminated. Review your
code to see if there are any unnecessary loops or computations.
............
```

ConversationBufferWindowMemory

ConversationBufferWindowMemory stores only the last "N" interactions, essentially creating a rolling window of recent conversation context. Unlike ConversationBufferMemory, this approach retains only the most recent exchanges, thereby reducing the storage burden.

- **Use Case:** Ideal for scenarios where only the latest context is relevant, such as chat-based Q&A or short-session customer support. It's also well-suited for lightweight applications where continuity is needed but only over recent exchanges.
- **Advantages:** It conserves resources by limiting the memory scope, which is useful for applications handling high volumes of user interactions.
- **Limitations:** Since it only keeps a limited number of exchanges, this memory type may lose earlier parts of the conversation, which could affect continuity in applications where longer context is essential.

Example:

```
from langchain.chat_models import ChatOpenAI
from langchain.prompts import ChatPromptTemplate,
SystemMessagePromptTemplate, HumanMessagePromptTemplate
from langchain.chains import LLMChain
from langchain.memory import ConversationBufferWindowMemory

# Initialize the chat model
chat_model = ChatOpenAI(model="gpt-3.5-turbo", temperature=0.7)

# Define the prompt templates
system_prompt = SystemMessagePromptTemplate.from_template("You are a
helpful assistant.")
human_prompt = HumanMessagePromptTemplate.from_template("{history}\n\nUser:
{input}")

# Wrap prompts in a ChatPromptTemplate
chat_prompt = ChatPromptTemplate.from_messages([system_prompt, human_
prompt])

# Set up the memory with a window of 3 messages
memory = ConversationBufferWindowMemory(k=3, return_messages=True)

# Create the chain with memory
conversation_chain = LLMChain(
    llm=chat_model,
```

```
    prompt=chat_prompt,
    memory=memory
)

# Example interactions
interactions = [
    "Hello, can you help me with some Python code?",
    "I need help with writing a loop.",
    "What are some best practices for functions?",
    "How do I make my code run faster?",
    "What should I know about error handling?",
]

# Running each interaction and printing the results, focusing on
memory usage
for i, user_input in enumerate(interactions, 1):
    print(f"Interaction {i}: User Input: {user_input}")
    response = conversation_chain.run(input=user_input)
    print(f"Assistant Response: {response}")

    # Print the current state of memory (only the last k interactions)
    current_memory = memory.load_memory_variables({})['history']
    memory_contents = [msg.content for msg in current_memory]
    print(f"Current Memory State: {memory_contents}\n")
```

Output:

```
Interaction 1: User Input: Hello, can you help me with some Python code?
Assistant Response: Of course! I'd be happy to help. What do you need
assistance with in Python?
Current Memory State: ['Hello, can you help me with some Python code?', "Of
course! I'd be happy to help. What do you need assistance with in Python?"]

Interaction 2: User Input: I need help with writing a loop.
Assistant Response: Sure! I can help with that. What specific task or
purpose would you like the loop to achieve?
Current Memory State: ['Hello, can you help me with some Python code?', "Of
course! I'd be happy to help. What do you need assistance with in Python?",
```

```
'I need help with writing a loop.', 'Sure! I can help with that. What
specific task or purpose would you like the loop to achieve?']

Interaction 3: User Input: What are some best practices for functions?
Assistant Response: When writing functions in Python, here are some best
practices to keep in mind:
1. **Function Naming**: Choose descriptive and meaningful names for your
functions that reflect their purpose or behavior. Use lowercase letters and
underscores to separate words (snake_case).
2. **Function Length**: Keep your functions concise and focused on a single
task. If a function becomes too long or complex, consider refactoring it
into smaller, more manageable functions…............
Current Memory State: ['Hello, can you help me with some Python code?',
"Of course! I'd be happy to help. What do you need assistance with in
Python?", 'I need help with writing a loop.', 'Sure! I can help with
that. What specific task or purpose would you like the loop to achieve?',
'What are some best practices for functions?', "When writing functions in
Python, here are some best practices to keep in mind:\n\n1. **Function
Naming**: Choose descriptive and meaningful names for your functions that
reflect their purpose or behavior. Use lowercase letters and underscores
to separate words (snake_case).\n\n2. **Function Length**: Keep your
functions concise and focused on a single task. If a function becomes too
long or complex, consider refactoring it into smaller, more manageable
functions.\n\n3. **Function Documentation**: Always include a docstring
at the beginning of your function to describe its purpose, parameters,
and return value. This helps other developers (and your future self)
understand the function's functionality.\n\n4. **Parameter Passing**: Avoid
using global variables inside functions. Instead, pass necessary data as
parameters to make your functions more modular and reusable.\n\n5. **Return
Values**: Clearly define what your function should return, even if it's
None. This makes it easier to understand how the function interacts with
the rest of your code.\n\n6. **Error Handling**: Implement appropriate
error handling within your functions to gracefully handle unexpected
situations and provide meaningful error messages to the user.\n\n7. **Code
Readability**: Write clean and readable code by following Python's style
guide (PEP 8). Use proper indentation, spacing, and naming conventions
```

```
to enhance code clarity.\n\n8. **Testing**: Write unit tests for your functions to ensure they work as expected under different scenarios. This helps catch bugs early and maintain the reliability of your code.\n\nBy following these best practices, you can write more maintainable, reusable, and understandable functions in your Python code."]

Interaction 4: User Input: How do I make my code run faster?
Assistant Response: Improving the performance of your code can involve various strategies. Here are some general tips to help make your Python code run faster:
1. **Use Efficient Data Structures**: Choose the appropriate data structures for your tasks. For example, use dictionaries for fast lookups, sets for membership tests, and lists for sequential data.
2. **Avoid Unnecessary Loops**: Minimize the number of loops and iterations in your code. Consider using list comprehensions or built-in functions like `map()`, `filter()`, and `reduce()`.........
Current Memory State: ['I need help with writing a loop.', 'Sure! I can help with that. What specific task or purpose would you like the loop to achieve?', 'What are some best practices for functions?', "When writing functions in Python, here are some best practices to keep in mind:\n\n1. **Function Naming**: Choose descriptive and meaningful names for your functions that reflect their purpose or behavior. Use lowercase letters and underscores to separate words (snake_case).\n\n2. **Function Length**: Keep your functions concise and focused on a single task. If a function becomes too long or complex, consider refactoring it into smaller, more manageable functions.\n\n3. **Function Documentation**: Always include a docstring at the beginning of your function to describe its purpose, parameters, and return value. This helps other developers (and your future self) understand the function's functionality.\n\n4. **Parameter Passing**: Avoid using global variables inside functions. Instead, pass necessary data as parameters to make your functions more modular and reusable.\n\n5. **Return Values**: Clearly define what your function should return, even if it's None. This makes it easier to understand how the function interacts with the rest of your code.\n\n6. **Error Handling**: Implement appropriate error handling within your functions to gracefully handle unexpected situations and provide meaningful error messages to the user.\n\
```

```
n7. **Code Readability**: Write clean and readable code by following
Python's style guide (PEP 8). Use proper indentation, spacing, and naming
conventions to enhance code clarity.\n\n8. **Testing**: Write unit tests
for your functions to ensure they work as expected under different
scenarios. This helps catch bugs early and maintain the reliability of
your code.\n\nBy following these best practices, you can write more
maintainable, reusable, and understandable functions in your Python code.",
'How do I make my code run faster?', "Improving the performance of your
code can involve various strategies. Here are some general tips to help
make your Python code run faster:\n\n1. **Use Efficient Data Structures**:
Choose the appropriate data structures for your tasks. For example, use
dictionaries for fast lookups, sets for membership tests, and lists for
sequential data.\n\n2. **Avoid Unnecessary Loops**: Minimize the number
of loops and iterations in your code. Consider using list comprehensions
or built-in functions like `map()`, `filter()`, and `reduce()`.\n\n3.
**Optimize Algorithm Complexity**: Analyze the algorithmic complexity
of your code and try to optimize it. Use efficient algorithms and data
structures to reduce time complexity.\n\n4. **Cache Results**: If certain
calculations or operations are repeated, consider caching the results
to avoid redundant computations.\n\n5. **Use Built-in Functions**: Take
advantage of Python's built-in functions and libraries, as they are often
optimized for performance.\n\n6. **Avoid Global Variables**: Minimize the
use of global variables, as accessing them can be slower compared to local
variables.\n\n7. **Profile Your Code**: Use Python's built-in profiling
tools like cProfile to identify bottlenecks in your code and optimize
the performance-critical sections.\n\n8. **Consider Cython or Numba**:
For computationally intensive tasks, consider using Cython or Numba to
compile Python code to C or machine code for improved performance.\n\
n9. **Optimize I/O Operations**: If your code involves reading or writing
large amounts of data, optimize I/O operations by using buffered I/O or
asynchronous programming.\n\n10. **Parallelize Tasks**: For tasks that
can be parallelized, consider using libraries like `multiprocessing`
or `concurrent.futures` to leverage multiple CPU cores for faster
execution.\n\nBy applying these strategies and considering the specific
requirements of your code, you can optimize its performance and make it run
faster."]
```

```
Interaction 5: User Input: What should I know about error handling?
Assistant Response: When it comes to error handling in Python, here are
some key points to keep in mind:

1. **Types of Errors**: Understand the different types of errors that can
occur in your code, such as syntax errors, runtime errors, and logical
errors. Python provides built-in exception classes to handle these errors.
2. **try-except Block**: Use a `try-except` block to catch and handle
exceptions in your code. The `try` block contains the code that might
raise an exception, while the `except` block handles the exception if it
occurs............
Current Memory State: ['What are some best practices for functions?',
"When writing functions in Python, here are some best practices to keep in
mind:\n\n1. **Function Naming**....."]
```

ConversationSummaryMemory

ConversationSummaryMemory creates a running summary of the conversation, synthesizing essential points while filtering out less relevant details. This summarized memory offers a condensed view of the interaction, capturing the conversation's main ideas, key decisions, and any important details that need continuity.

- **Use Case:** Suitable for applications requiring ongoing context without an exhaustive log, such as personal assistants, tutoring systems, or patient tracking in healthcare, where high-level summaries of conversations provide value.
- **Advantages:** Reduces memory load by storing only summarized data while retaining crucial context. This balance between detail and memory efficiency helps create responses that maintain coherence without overwhelming resources.
- **Limitations:** Summarization may overlook nuances or less prominent details, which could be essential for some applications. Developing an effective summarization approach is critical for making this memory type work well.

Example:

```
from langchain.chat_models import ChatOpenAI
from langchain.prompts import ChatPromptTemplate,
SystemMessagePromptTemplate, HumanMessagePromptTemplate
from langchain.chains import LLMChain
from langchain.memory import ConversationSummaryMemory

# Initialize the chat model
chat_model = ChatOpenAI(model="gpt-3.5-turbo", temperature=0.7)

# Define the prompt templates
system_prompt = SystemMessagePromptTemplate.from_template("You are a
helpful assistant.")
human_prompt = HumanMessagePromptTemplate.from_template("{history}\n\nUser:
{input}")

# Wrap prompts in a ChatPromptTemplate
chat_prompt = ChatPromptTemplate.from_messages([system_prompt, human_
prompt])

# Set up the summary memory
memory = ConversationSummaryMemory(llm=chat_model)

# Create the chain with memory
conversation_chain = LLMChain(
    llm=chat_model,
    prompt=chat_prompt,
    memory=memory
)

# Example interactions
interactions = [
    "Hello, can you help me with some Python code?",
    "I need help with writing a loop."
]

# Running each interaction and printing the results, focusing on
memory usage
```

```
for i, user_input in enumerate(interactions, 1):
    print(f"Interaction {i}: User Input: {user_input}")
    response = conversation_chain.run(input=user_input)
    print(f"Assistant Response: {response}")

    # Print the current summarized state of memory
    current_summary = memory.load_memory_variables({})['history']
    print(f"Current Memory Summary: {current_summary}\n")
```

Output:

```
Interaction 1: User Input: Hello, can you help me with some Python code?
Assistant Response: Sure, I'd be happy to help! What specifically do you
need assistance with in your Python code?
Current Memory Summary: The human asks the AI for help with some Python
code. The AI is willing to assist and asks for specifics about the code
that needs help.
Interaction 2: User Input: I need help with writing a loop.
Assistant Response: Of course! I'd be happy to help. Could you please
provide more details about what you are trying to achieve with the loop in
your Python code?
Current Memory Summary: The human asks the AI for help with some Python
code. The AI is willing to assist and asks for specifics about the code
that needs help, such as details about the loop the human is trying
to write.
```

Conversation Summary Buffer Memory

Conversation Summary Buffer Memory combines conversation summarization with a recent message buffer, offering a memory that captures both high-level context and recent details. This approach is useful for applications that need to retain the essence of previous exchanges while keeping immediate context close at hand.

- **How It Works**: Conversation Summary Buffer Memory continuously updates a summary of the conversation's main ideas, storing the most essential points from past interactions. Alongside this summary, it maintains a small, recent buffer containing the last "N" messages in full, providing immediate context without overloading the memory with the entire conversation history.

- **Use Case**: Ideal for applications requiring a mix of long-term continuity and immediate context, such as personal assistants, coaching applications, or educational tools. For example, a therapeutic chatbot might retain a summary of past sessions while also keeping recent exchanges to stay relevant in ongoing dialogues.
- **Advantages**: This memory type balances memory usage and contextual richness, keeping a distilled summary to capture key points over time while retaining recent details. This design ensures the model can respond with continuity and relevance, even across extended conversations.
- **Limitations**: Summarization may miss minor details or nuances not captured in the main summary, which could affect applications needing precise recall of historical interactions. Additionally, the limited buffer size means that only a small portion of recent exchanges is retained verbatim, which may not suit applications that need a longer-term message history.

Example:

```
from langchain.chat_models import ChatOpenAI
from langchain.prompts import ChatPromptTemplate,
SystemMessagePromptTemplate, HumanMessagePromptTemplate
from langchain.chains import LLMChain
from langchain.memory import ConversationSummaryBufferMemory

# Initialize the chat model
chat_model = ChatOpenAI(model="gpt-3.5-turbo", temperature=0.7)

# Define the prompt templates
system_prompt = SystemMessagePromptTemplate.from_template("You are a
helpful assistant.")
human_prompt = HumanMessagePromptTemplate.from_template("{history}\n\nUser:
{input}")
```

```
# Wrap prompts in a ChatPromptTemplate
chat_prompt = ChatPromptTemplate.from_messages([system_prompt, human_
prompt])

# Set up the summary buffer memory with a window of 3 messages
memory = ConversationSummaryBufferMemory(llm=chat_model, max_token_
limit=100)

# Create the chain with memory
conversation_chain = LLMChain(
    llm=chat_model,
    prompt=chat_prompt,
    memory=memory
)

# Example interactions
interactions = [
    "Hello, can you help me with some Python code?",
    "I need help with writing a loop."
]

# Running each interaction and printing the results, focusing on
memory usage
for i, user_input in enumerate(interactions, 1):
    print(f"Interaction {i}: User Input: {user_input}")
    response = conversation_chain.run(input=user_input)
    print(f"Assistant Response: {response}")

    # Print the current summarized state of memory
    current_summary = memory.load_memory_variables({})['history']
    print(f"Current Memory Summary and Recent Buffer: {current_summary}\n")
```

Example:

```
Interaction 1: User Input: Hello, can you help me with some Python code?
Assistant Response: Of course! I'll do my best to help you with your Python
code. What do you need assistance with?
Current Memory Summary and Recent Buffer: Human: Hello, can you help me
with some Python code?
```

```
AI: Of course! I'll do my best to help you with your Python code. What do
you need assistance with?

Interaction 2: User Input: I need help with writing a loop.
Assistant Response: AI: Sure, I'd be happy to help you with writing a
loop in Python. What specific task or goal would you like to achieve with
the loop?
Current Memory Summary and Recent Buffer: Human: Hello, can you help me
with some Python code?
AI: Of course! I'll do my best to help you with your Python code. What do
you need assistance with?
Human: I need help with writing a loop.
AI: AI: Sure, I'd be happy to help you with writing a loop in Python. What
specific task or goal would you like to achieve with the loop?
```

Conversation Token Buffer Memory

Conversation Token Buffer Memory manages conversation memory based on a defined token limit, maintaining recent exchanges within a specified token capacity rather than message count. This memory type ensures efficient context retention by storing interactions until a preset token threshold is reached.

- **How It Works**: Conversation Token Buffer Memory tracks the number of tokens used in each message and trims older interactions as new ones are added once the token count exceeds the set limit. By using tokens as the metric, this memory type accommodates messages of varying lengths without exceeding model processing limits.
- **Use Case**: Well-suited for applications needing to stay within strict token constraints, such as chatbots with token-based memory limits or customer service agents operating within resource constraints. For instance, a support chatbot could retain recent exchanges within a token cap, ensuring the model remains within processing capacity while preserving relevant context.

- **Advantages**: The token-based approach offers flexibility and efficiency, especially for applications with limited memory resources. This memory type adapts easily to conversations with variable message lengths, ensuring that the most recent context is preserved within a fixed token boundary.
- **Limitations**: Important context may be lost when older messages are removed due to token limits, which could affect continuity in longer conversations. Additionally, shorter messages may result in more frequent trimming if the token cap is low, potentially impacting continuity in extensive sessions.

Example:

```
from langchain.chat_models import ChatOpenAI
from langchain.prompts import ChatPromptTemplate,
SystemMessagePromptTemplate, HumanMessagePromptTemplate
from langchain.chains import LLMChain
from langchain.memory import ConversationTokenBufferMemory

# Initialize the chat model
chat_model = ChatOpenAI(model="gpt-3.5-turbo", temperature=0.7)

# Define the prompt templates
system_prompt = SystemMessagePromptTemplate.from_template("You are a
helpful assistant.")
human_prompt = HumanMessagePromptTemplate.from_template("{history}\n\nUser:
{input}")

# Wrap prompts in a ChatPromptTemplate
chat_prompt = ChatPromptTemplate.from_messages([system_prompt, human_
prompt])

# Set up the token buffer memory with a small max token limit
memory = ConversationTokenBufferMemory(llm=chat_model, max_token_limit=30)

# Create the chain with memory
conversation_chain = LLMChain(
    llm=chat_model,
    prompt=chat_prompt,
```

```
    memory=memory
)

# Example interactions
interactions = [
    "Hi, I need help with Python.",
    "How do I create a list?"
]

# Run each interaction and print the memory state
for i, user_input in enumerate(interactions, 1):
    print(f"Interaction {i}: User Input: {user_input}")
    response = conversation_chain.run(input=user_input)
    print(f"Assistant Response: {response}")

    # Print the current state of the token-limited memory buffer
    current_buffer = memory.load_memory_variables({})['history']
    print(f"Current Memory Buffer (Token-Limited): {current_buffer}\n")
```

Output:

````
Interaction 1: User Input: Hi, I need help with Python.
Assistant Response: Sure, I'd be happy to help you with Python. What
specifically do you need assistance with?
Current Memory Buffer (Token-Limited): AI: Sure, I'd be happy to help you
with Python. What specifically do you need assistance with?

Interaction 2: User Input: How do I create a list?
Assistant Response: To create a list in Python, you can use square brackets
`[ ]` and separate the elements with commas. Here's an example:

```python
my_list = [1, 2, 3, 4, 5]
```

In this example, `my_list` is a list containing the numbers 1, 2, 3, 4, and
5. You can create lists with different types of elements, such as strings,
numbers, or even other lists. Let me know if you need further assistance
with lists or any other Python topics!
Current Memory Buffer (Token-Limited):
````

KnowledgeGraphMemory

KnowledgeGraphMemory organizes information into a knowledge graph structure, representing interactions in terms of entities and their relationships. By storing and linking entities, KnowledgeGraphMemory allows the model to reference and reason about relationships, creating a structured representation of the conversation history.

- **Use Case:** This memory type excels in applications where complex relationships or structured knowledge is critical, such as recommendation engines, expert systems, or domain-specific assistants (e.g., legal or medical applications). It's also useful for situations requiring relational understanding.
- **Advantages:** Provides a structured and rich contextual layer, allowing the model to perform entity-based reasoning and relational analysis. It enables advanced interactions where understanding relationships is crucial.
- **Limitations:** Implementing a knowledge graph structure can be more complex and computationally intensive than simpler memory types, especially as the number of entities and relationships grows.

Example:

```
from langchain.chat_models import ChatOpenAI
from langchain.prompts import ChatPromptTemplate,
SystemMessagePromptTemplate, HumanMessagePromptTemplate
from langchain.chains import LLMChain
from langchain.memory import ConversationKGMemory  # Ensure this class is
available in your version

# Initialize the chat model
chat_model = ChatOpenAI(model="gpt-3.5-turbo", temperature=0.7)

# Define the prompt templates
system_prompt = SystemMessagePromptTemplate.from_template("You are a
helpful assistant that keeps track of information in a knowledge graph.")
human_prompt = HumanMessagePromptTemplate.from_template("User: {input}")
```

```
# Wrap prompts in a ChatPromptTemplate with an expected input variable
chat_prompt = ChatPromptTemplate.from_messages([system_prompt, human_
prompt])

# Set up the Knowledge Graph memory
memory = ConversationKGMemory(llm=chat_model)

# Create the chain with memory
conversation_chain = LLMChain(
    llm=chat_model,
    prompt=chat_prompt,
    memory=memory
)

# Example interactions
interactions = [
    "Alice is a software engineer.",
    "Alice works at OpenAI.",
    "Bob is Alice's manager.",
]

# Run each interaction and print the knowledge graph state
for i, user_input in enumerate(interactions, 1):
    print(f"Interaction {i}: User Input: {user_input}")
    response = conversation_chain.run(input=user_input)
    # Passing the user input
    print(f"Assistant Response: {response}")

    # Retrieve and print all memory variables to check for the
      knowledge graph
    try:
        memory_variables = memory.load_memory_variables({"input": "who is
        bob"})  # Provide a dummy input
        print("Memory Variables:", memory_variables)

    except ValueError as e:
        print(f"Error retrieving memory variables: {e}")
```

Output:

```
Interaction 1: User Input: Alice is a software engineer.
Assistant Response: Statement added: Alice is a software engineer.
Memory Variables: {'history': 'On Alice: Alice is a software engineer.'}
Interaction 2: User Input: Alice works at OpenAI.
Assistant Response: Got it! Alice works at OpenAI.
Memory Variables: {'history': ''}
Interaction 3: User Input: Bob is Alice's manager.
Assistant Response: Statement recorded: Bob is Alice's manager.
Memory Variables: {'history': 'On Bob: Bob is manager. Bob manages Alice.'}
```

EntityMemory

EntityMemory focuses specifically on tracking entities and their relevant details across interactions. Rather than storing the full transcript, EntityMemory captures only specific attributes or facts associated with key entities, such as a user's name, preferences, or recurring topics.

- **Use Case:** This memory type is particularly valuable for applications centered around user profiles or personalization, such as customer service chatbots, ecommerce recommendation systems, and user-focused applications where remembering specific details about users enhances the experience.
- **Advantages:** By honing in on relevant entities, EntityMemory helps the model recall user-specific details efficiently, without the overhead of full conversations. It's an effective way to provide a personalized experience without excessive memory usage.
- **Limitations:** Since only targeted entity information is stored, this memory type might miss broader context outside of the entity-specific data, which could affect applications needing holistic continuity across sessions.

Example:

```
from langchain.chat_models import ChatOpenAI
from langchain.prompts import ChatPromptTemplate,
SystemMessagePromptTemplate, HumanMessagePromptTemplate
from langchain.chains import LLMChain
from langchain.memory import ConversationEntityMemory

# Initialize the chat model
chat_model = ChatOpenAI(model="gpt-3.5-turbo", temperature=0.7)

# Define the prompt templates
system_prompt = SystemMessagePromptTemplate.from_template("You are
a helpful assistant that keeps track of entities mentioned in the
conversation.")
human_prompt = HumanMessagePromptTemplate.from_template("User: {input}\n")

# Wrap prompts in a ChatPromptTemplate
chat_prompt = ChatPromptTemplate.from_messages([system_prompt, human_
prompt])

# Set up the Entity Memory
memory = ConversationEntityMemory(llm=chat_model)

# Create the chain with memory
conversation_chain = LLMChain(
    llm=chat_model,
    prompt=chat_prompt,
    memory=memory
)

# Example interactions
interactions = [
    "Alice is a software engineer.",
    "Alice works at OpenAI.",
    "Bob is Alice's manager.",
]

# Run each interaction and print the entity memory state
```

```
for i, user_input in enumerate(interactions, 1):
    print(f"Interaction {i}: User Input: {user_input}")
    response = conversation_chain.run(input=user_input)
    print(f"Assistant Response: {response}")

    # Retrieve and print the current state of entity memory
    entity_memory = memory.load_memory_variables({"input": "Show me
    names?"}).get("entities", "No entities tracked.")
    print(f"Current Entity Memory: {entity_memory}\n")
```

Output:

```
Interaction 1: User Input: Alice is a software engineer.
Assistant Response: Got it! Alice is a software engineer.
Current Entity Memory: {'Alice': 'Alice is a software engineer.'}

Interaction 2: User Input: Alice works at OpenAI.
Assistant Response: Got it! Alice works at OpenAI.
Current Entity Memory: {'Alice': 'Alice is a software engineer who works at
OpenAI.', 'OpenAI': 'OpenAI is the workplace of Alice.'}

Interaction 3: User Input: Bob is Alice's manager.
Assistant Response: Got it! Bob is Alice's manager.
Current Entity Memory: {'Alice': 'Alice is a software engineer who works at
OpenAI, and Bob is her manager.', 'OpenAI': 'OpenAI is where Alice works,
and Bob is her manager.', 'Bob': "Bob is Alice's manager."}
```

VectorStoreMemory

VectorStoreMemory employs vector embeddings to store conversation elements based on semantic similarity, rather than literal text. By encoding past interactions into vector space, this memory type enables retrieval based on conceptual similarity, allowing the model to identify relevant topics or contexts from previous conversations.

- **Use Case:** This approach is ideal for applications needing rapid access to related information, such as personalized content recommendations, topic-based knowledge retrieval, and advanced conversational systems that require nuanced understanding of user context over time.

- **Advantages:** Vector-based memory enables flexible and rapid retrieval, allowing the model to match current queries with semantically similar past interactions. This enhances contextual relevance, especially in applications where users revisit similar topics or inquiries.
- **Limitations:** Vector-based memory requires computational resources to compute and store embeddings. Additionally, the retrieval may sometimes favor conceptually similar information over exact conversational details, which could affect applications where exact recall is needed.

Example:

```
from langchain.chat_models import ChatOpenAI
from langchain.prompts import ChatPromptTemplate,
SystemMessagePromptTemplate, HumanMessagePromptTemplate
from langchain.chains import LLMChain
from langchain.memory import VectorStoreRetrieverMemory
from langchain.vectorstores import FAISS
from langchain.embeddings import OpenAIEmbeddings
from langchain.docstore import InMemoryDocstore
import faiss
import numpy as np

# Initialize the chat model
chat_model = ChatOpenAI(model="gpt-3.5-turbo", temperature=0.7)

# Initialize the embedding model for vector storage
embedding_model = OpenAIEmbeddings()

# Set up FAISS index with the correct embedding dimension
embedding_dim = 1536  # Ensure this matches the dimension of embeddings
index = faiss.IndexFlatL2(embedding_dim)

# Set up FAISS vector store with additional required components
vector_store = FAISS(
    embedding_function=embedding_model.embed_query,
    index=index,
```

```
    docstore=InMemoryDocstore({}),  # Initialize an empty docstore
    index_to_docstore_id={}          # Start with an empty ID mapping
)

# Define the prompt templates
system_prompt = SystemMessagePromptTemplate.from_template("You are a
helpful assistant with memory capabilities.")
human_prompt = HumanMessagePromptTemplate.from_template("{history}\n\nUser:
{input}")

# Wrap prompts in a ChatPromptTemplate
chat_prompt = ChatPromptTemplate.from_messages([system_prompt, human_
prompt])

# Set up VectorStoreRetrieverMemory with the FAISS vector store
memory = VectorStoreRetrieverMemory(retriever=vector_store.as_retriever())

# Create the chain with memory
conversation_chain = LLMChain(
    llm=chat_model,
    prompt=chat_prompt,
    memory=memory
)

# Example interactions to store in memory
interactions = [
    "I'm planning a trip to Italy.",
    "Can you suggest some historic sites to visit?"
    ]

# Run each interaction, store it in the vector memory, and display
retrievals
for i, user_input in enumerate(interactions, 1):
    print(f"Interaction {i}: User Input: {user_input}")
    response = conversation_chain.run(input=user_input)
    print(f"Assistant Response: {response}")
```

```
    # Retrieve similar memory entries based on the latest user input
    related_memory = memory.retriever.get_relevant_documents(user_input)
    print("\nRelated Memory Entries (from VectorStore):")
    for entry in related_memory:
        print(f"- {entry.page_content}")
    print("\n" + "="*50 + "\n")
```

Output:

```
Interaction 1: User Input: I'm planning a trip to Italy.
Assistant Response: That's great! Italy is a beautiful country with so
much to see and do. Do you need any help with planning your trip or
recommendations on places to visit?
Related Memory Entries (from VectorStore):
- input: I'm planning a trip to Italy.
text: That's great! Italy is a beautiful country with so much to see and
do. Do you need any help with planning your trip or recommendations on
places to visit?
==================================================
Interaction 2: User Input: Can you suggest some historic sites to visit?
Assistant Response: Sure! Italy is full of historic sites that are
definitely worth visiting. Here are some popular historic sites in Italy:
1. The Colosseum in Rome: A iconic symbol of ancient Rome, this
amphitheater is one of the most well-preserved Roman structures in
the world.
2. The Roman Forum in Rome: Once the center of Roman public life, the Roman
Forum is a sprawling archaeological site with ruins of ancient government
buildings, temples, and monuments.
3. Pompeii: This ancient Roman city was buried by the eruption of Mount
Vesuvius in 79 AD, preserving it in remarkable detail. You can explore the
well-preserved ruins of homes, temples, and public buildings.
4. The Leaning Tower of Pisa: Located in the city of Pisa, this iconic
tower is known for its distinctive lean and is part of the Cathedral Square
complex, a UNESCO World Heritage Site.
```

```
5. The Vatican City: A city-state within Rome, the Vatican is home to St.
Peter's Basilica, the Sistine Chapel, and the Vatican Museums, which house
an incredible collection of art and artifacts.
These are just a few of the many historic sites you can visit in Italy. Let
me know if you need more information or recommendations!

Related Memory Entries (from VectorStore):
- input: Can you suggest some historic sites to visit?
text: Sure! Italy is full of historic sites that are definitely worth
visiting. Here are some popular historic sites in Italy:
1. The Colosseum in Rome: A iconic symbol of ancient Rome, this
amphitheater is one of the most well-preserved Roman structures in
the world.
2. The Roman Forum in Rome: Once the center of Roman public life, the Roman
Forum is a sprawling archaeological site with ruins of ancient government
buildings, temples, and monuments.
3. Pompeii: This ancient Roman city was buried by the eruption of Mount
Vesuvius in 79 AD, preserving it in remarkable detail. You can explore the
well-preserved ruins of homes, temples, and public buildings.
4. The Leaning Tower of Pisa: Located in the city of Pisa, this iconic
tower is known for its distinctive lean and is part of the Cathedral Square
complex, a UNESCO World Heritage Site.
5. The Vatican City: A city-state within Rome, the Vatican is home to St.
Peter's Basilica, the Sistine Chapel, and the Vatican Museums, which house
an incredible collection of art and artifacts.
These are just a few of the many historic sites you can visit in Italy. Let
me know if you need more information or recommendations!
- input: I'm planning a trip to Italy.
text: That's great! Italy is a beautiful country with so much to see and
do. Do you need any help with planning your trip or recommendations on
places to visit?
==========================================================================
```

Selecting the Appropriate Memory Type

Each LangChain Memory type has unique strengths, tailored to specific application needs:

- **Full Conversation Memory** (ConversationBufferMemory): For applications needing complete historical context
- **Recent Context Memory** (ConversationBufferWindowMemory): For lightweight interactions focused on immediate past exchanges
- **Summarized Context Memory** (ConversationSummaryMemory): For condensed, high-level overviews that retain key points without storing details
- **Relational Memory** (KnowledgeGraphMemory): For structured applications needing entity-based reasoning and complex relationship tracking
- **Entity-Focused Memory** (EntityMemory): For personalized user experiences based on key details about entities
- **Semantic Memory** (VectorStoreMemory): For flexible, concept-driven applications that benefit from semantic similarity retrieval

Implementing Memory in LangChain

To utilize memory effectively within LangChain, developers need to configure the appropriate memory type based on the application's requirements. Each type can be initialized with specific settings that determine how data is stored, retrieved, and utilized within a conversational chain. Implementing memory in LangChain typically involves

- **Setting Up the Memory Type:** Initialize the selected memory type and configure its storage limits, retrieval logic, and any other relevant parameters.
- **Integrating with the LLMChain:** Embed the memory module into the LLMChain to ensure the model reads from memory during each call, supplements user inputs with contextual information, and writes new data back to memory.

- **Managing Memory Life Cycle:** Depending on the application, developers may need to define how long information persists in memory and establish any clearing or summarizing protocols as interactions grow.

LangChain's memory options provide both flexibility and precision, empowering developers to create conversational AI systems that adapt seamlessly to user needs, preserve relevant context, and enhance the coherence of interactions. By selecting the memory type that best aligns with the application's goals, developers can build more responsive, personalized, and efficient conversational agents.

When deploying a LangChain-powered RAG (retrieval-augmented generation) server in production—especially in environments where multiple replicas or pods are used, such as in Kubernetes—it is crucial to architect the memory system in a way that ensures scalability, consistency, and persistence across instances. LangChain's memory features are powerful for enabling continuity in conversational AI, but in a stateless deployment model, developers must externalize memory storage to avoid context loss or inconsistency between user sessions.

To persist memory across replicas, it is recommended to use an external memory store. Options include Redis (e.g., via RedisChatMessageHistory); vector databases like FAISS, Pinecone, Weaviate, or Qdrant for semantic memory; or even traditional databases such as PostgreSQL or MongoDB for storing structured conversation logs. These solutions allow all replicas to read from and write to a centralized memory source, ensuring that user sessions remain consistent regardless of which pod processes the request.

Each user or conversation should be associated with a unique session identifier—such as a user_id, session_id, or conversation_id—which should be passed with every request. This allows the server to correctly retrieve the corresponding memory from the centralized store. Middleware in your API layer can be used to manage session resolution and memory access, ensuring the right context is injected into each interaction with the language model.

While sticky sessions can be used temporarily to route users to the same pod, this approach is not recommended for long-term scalability or reliability. Centralized memory storage is a more robust solution, especially in distributed environments that may auto-scale or experience pod restarts.

Memory management also involves defining life cycle rules. Developers should use TTL (Time-To-Live) settings or similar expiration mechanisms to automatically clean up unused or stale memory entries. LangChain also supports summarization strategies that help reduce memory size over time while retaining core contextual information. Applications should implement logic to determine when memory should be reset or archived, such as after a defined period of inactivity or at the end of a conversation session.

From an operational perspective, logging and monitoring memory access is essential. Developers should track memory reads, writes, and retrieval times and set up alerts for failures or inconsistencies. Tools such as Prometheus, Grafana, and OpenTelemetry can provide visibility into memory performance and help detect anomalies.

For multitenant applications, memory should be logically partitioned by tenant_id to ensure data isolation and compliance. Role-based access control should be enforced on the memory back end to prevent unauthorized access across tenants or sessions.

Prior to production rollout, load testing should be conducted to evaluate memory pressure under concurrent sessions and long conversations. Testing should verify that memory retrieval is performant and that API token usage remains within budget, especially if memory content is included in each prompt sent to the LLM.

In summary, production deployment of memory-enabled RAG servers using LangChain requires careful planning and infrastructure. By externalizing memory storage, implementing robust session management, and monitoring life cycle and performance, developers can build scalable, reliable, and context-aware conversational systems that maintain coherence across distributed workloads.

LangChain Document Loaders

LangChain's *Document Loaders* simplify data ingestion by offering flexible, modular ways to load and preprocess data from a wide array of file types and online sources. With document loaders, users can retrieve structured and unstructured data from formats such as PDFs, Word documents, web pages, and even APIs, making it easier to analyze, summarize, or use the content within machine learning applications. This versatility in data sourcing is crucial for applications involving document search, question answering, or content generation, where the quality of data input directly impacts results.

Here is an expanded list of some of the popular and specialized document loaders supported by LangChain, as noted in the LangChain documentation.

Common Document Loaders

- **PDF Loaders**: Extracts text from PDF files and supports different parsing methods, making it suitable for documents that include images, tables, or specific layouts.
- **Word Document Loader**: This loader extracts text from `.doc` and `.docx` formats, making it ideal for processing Microsoft Word documents.
- **CSV and Excel Loaders**: These loaders bring in data from CSV and Excel files, organizing tabular data that can be beneficial for structured datasets, reports, and analytics workflows.
- **Notion Loader**: Enables direct integration with Notion, allowing users to pull data from Notion pages and databases for teams that work collaboratively in this platform.
- **Web Page Loader**: A versatile loader that fetches content from URLs, transforming raw web pages into structured text suitable for processing in NLP applications.
- **Google Drive Loader**: This loader retrieves documents from Google Drive, making it easy to process cloud-stored files shared within organizations or teams.

Specialized Document Loaders

- **HTML Loader**: Imports data from HTML files, allowing users to capture structured web content in its native format, often useful for scraping and processing online content.
- **Markdown Loader**: Processes Markdown files, commonly used for documentation and technical content, to ensure compatibility in documentation-heavy workflows.

- **S3 and Azure Blob Storage Loaders**: Connect to cloud storage solutions like Amazon S3 and Azure Blob Storage, ideal for organizations with large datasets stored in these services.
- **Gmail Loader**: Retrieves emails from a Gmail account, parsing email threads for analysis, insights, or summarization.
- **YouTube Loader**: Allows loading transcripts and captions from YouTube videos, providing an easy way to turn video content into text for analysis or summarization.
- **API Loader**: A flexible loader that integrates with APIs, allowing users to bring in real-time data from external services or databases.
- **Slack Loader**: Loads message data from Slack, useful for teams needing to analyze conversations, gather feedback, or synthesize team communications.
- **Dropbox Loader**: Connects to Dropbox, enabling access to files stored in this platform and allowing data analysis for collaborative or cloud-based environments.
- **Confluence Loader**: Retrieves content from Confluence, useful for teams using it as their documentation or knowledge management tool.
- **GitHub Repository Loader**: Pulls text from GitHub repositories, useful for software documentation, code analysis, or processing README files and other documentation stored in GitHub.
- **RSS Feed Loader**: Loads data from RSS feeds, making it convenient for applications needing to stay updated with live information from news sites, blogs, or other sources.
- **JSON and XML Loaders**: These loaders are used for structured data in JSON and XML formats, common in APIs, data interchange formats, and various structured data applications.

By leveraging these document loaders, LangChain allows users to build sophisticated data pipelines tailored to the specific data needs of their NLP and machine learning workflows. The wide range of loaders supports various content sources, allowing teams to seamlessly integrate data from multiple platforms and formats into their applications.

Example:

```
from langchain.document_loaders import TextLoader
# Initialize the TextLoader with the path to the text file
file_path = "example.txt"  # Replace with your text file path
loader = TextLoader(file_path)

# Load the document
documents = loader.load()

# Display the loaded document
for doc in documents:
    print("Document Content:")
    print(doc.page_content)
    print("\nMetadata:")
    print(doc.metadata)
```

Output:

Depending on your txt file content.

If you don't have a document feel free to download one with dummy data by using this command: `curl https://sample-files.com/downloads/documents/txt/long-doc.txt> ./example.txt`

LangChain Embedding Models

An embedding model is a type of machine learning model that converts data, such as words, sentences, or images, into dense vector representations, often called *embeddings*. These embeddings are typically high-dimensional numeric arrays that capture the semantic or structural characteristics of the input data, allowing similar items to have similar vector representations. In natural language processing, embedding models enable computers to understand relationships and meanings between words or sentences by placing them in a continuous, multidimensional space where related concepts are closer together. This transformation facilitates tasks like search, classification, and similarity measurement by making comparisons between items more efficient and intuitive.

The field of embedding models has undergone substantial development over the years. A key turning point arrived in 2018 when Google launched BERT (Bidirectional Encoder Representations from Transformers), a model that transformed text into vector representations, achieving remarkable performance across numerous NLP tasks. Despite its advancements, BERT was not optimized for creating sentence embeddings efficiently, leading to the development of SBERT (Sentence-BERT).

SBERT adapted BERT's architecture to produce semantically rich sentence embeddings, which could be quickly compared using similarity metrics like cosine similarity, significantly reducing the computational demands for tasks such as sentence similarity searches. Today, the ecosystem of embedding models is varied, with many providers offering unique implementations. Researchers and practitioners frequently consult benchmarks like the Massive Text Embedding Benchmark (MTEB) for objective performance comparisons.

Unified Interface for Embedding Models

LangChain offers a standardized interface to interact with various embedding models, streamlining the process through two core methods:

- **embed_documents**: Embeds multiple texts (documents)
- **embed_query**: Embeds a single text (query)

This differentiation is essential, as some providers implement distinct embedding approaches for documents, which are used as searchable content, and for queries, which serve as the search input. For instance, LangChain's `.embed_documents` method can efficiently embed a list of text strings.

Measuring Similarity in Embedding Space

Each embedding acts as a coordinate in a high-dimensional space, where the position of each point reflects the meaning of its text. In this space, texts with similar meanings are located close to one another, akin to synonyms in a thesaurus. Converting text into numerical representations allows for swift similarity comparisons between text pairs, independent of their original form or length. Common similarity metrics include

- **Cosine Similarity**: Measures the cosine of the angle between two vectors
- **Euclidean Distance**: Calculates the straight-line distance between two points
- **Dot Product**: Determines the projection of one vector onto another

This approach enables meaningful and efficient comparisons, making it a foundational technique in modern NLP applications.

```
from langchain_openai import OpenAIEmbeddings
embeddings_model = OpenAIEmbeddings()
embeddings = embeddings_model.embed_documents(
    [
        "Hi there!",
        "Oh, hello!",
        "What's your name?",
        "My friends call me World",
        "Hello World!"
    ]
)
len(embeddings), len(embeddings[0])
```

Output:

```
(5, 1536)
```

LangChain integrates with a diverse array of embedding models, enabling users to generate vector representations of text for various applications. Here are 20 notable embedding models available within LangChain:

1. **OpenAI Embeddings**: Provides robust embeddings suitable for a wide range of natural language processing tasks

2. **Cohere Embeddings**: Offers versatile embeddings designed for tasks such as semantic search and text classification

3. **Hugging Face Transformers**: Features a collection of transformer-based models capable of producing high-quality embeddings for different languages and domains

4. **Google Vertex AI**: Delivers embeddings through Google's managed machine learning platform, facilitating seamless integration with other Google Cloud services

5. **Nomic Embeddings**: Specializes in embeddings tailored for large-scale data visualization and analysis

6. **IBM watsonx.ai**: Provides embeddings as part of IBM's suite of AI tools, suitable for enterprise applications
7. **Amazon Bedrock**: Offers embeddings through Amazon's fully managed service, supporting various AI applications
8. **DeepInfra Embeddings**: Utilizes serverless inference to provide access to a variety of LLMs and embedding models
9. **Jina AI**: Provides high-performance embeddings optimized for search and retrieval tasks
10. **GigaChat Embeddings**: Offers embeddings designed for conversational AI applications
11. **GPT4All**: A free-to-use, locally running chatbot that provides embeddings without requiring Internet access
12. **Gradient AI**: Allows creation of embeddings and fine-tuning of LLMs through a simple web API
13. **Fireworks Embeddings**: Provides embeddings included in the langchain_fireworks package for text embedding tasks
14. **Elasticsearch**: Generates embeddings using a hosted embedding model within the Elasticsearch platform
15. **ERNIE**: A text representation model based on Baidu Wenxin large-scale model technology
16. **FastEmbed by Qdrant**: A lightweight, fast Python library built for embedding generation
17. **LASER**: Language-Agnostic SEntence Representations by Meta AI, supporting multiple languages
18. **Llama-cpp**: Provides embeddings using the Llama-cpp library
19. **MiniMax**: Offers an embeddings service suitable for various NLP tasks
20. **MistralAI**: Provides embeddings through MistralAI's models, suitable for diverse applications

These integrations allow users to select the most appropriate embedding model for their specific needs, leveraging LangChain's unified interface to streamline the process.

LangChain Indexes and Retrievers

In LangChain, **indexes** and **retrievers** are essential tools that manage large datasets for applications using large language models (LLMs). These components are critical in efficiently storing and retrieving relevant information, powering applications like question-answering (QA) systems, chatbots, document search, and retrieval-augmented generation (RAG). Here's an in-depth look at how each of these components works.

Indexes in LangChain: Structure and Types

Indexes are data structures that organize and store datasets, making them accessible for quick retrieval. This process typically involves loading documents, breaking them into manageable chunks, embedding these chunks into vector representations, and creating indexes. Indexes in LangChain can be of various types, each designed to suit different application needs.

- **Document Loading and Chunking**: In the indexing process, documents are first loaded and divided into smaller chunks. Text splitters are used to create chunks that are small enough for efficient processing while retaining context. This chunking process is especially useful for handling large documents that exceed typical processing limits.
- **Embedding**: Each chunk of text is embedded into a high-dimensional vector space, where similar content resides near each other. This embedding step is crucial for vector indexes, which rely on semantic similarities between vectors to identify relevant information.
- **Index Creation**: LangChain's API offers a flexible approach to creating indexes, allowing developers to build different types of indexes based on their application needs. By utilizing embedding models, these indexes capture the nuances of each document, enabling advanced search capabilities across the dataset.

Types of Indexes

- **Vector Indexes**: Vector indexes convert document chunks into vectors that capture their semantic meaning. When a user query is converted into a vector, the vector index can perform similarity searches to retrieve documents close to the query in vector space. This approach is particularly useful for RAG applications, where contextual relevance is essential.
- **Keyword Indexes**: For applications focused on specific keywords, keyword indexes use sparse retrieval methods, such as term frequency-inverse document frequency (TF-IDF) or BM25, to match exact keywords in the documents. Though quicker than vector indexes, they are less capable of capturing the deeper semantic relationships between words.
- **Hybrid and Custom Indexes**: LangChain also supports hybrid indexes, which combine vector and keyword matching for applications requiring both semantic and exact keyword relevance. Custom indexes enable developers to define specialized retrieval logic, making them adaptable for domain-specific needs.

LangChain's indexing API is designed to be efficient, tracking document versions through hashing to ensure that only modified content is reindexed. This setup minimizes redundant data processing and maintains an up-to-date index.

Retrievers in LangChain: Querying and Optimization

Retrievers are components that query indexes to extract relevant document chunks based on a user query. They manage how indexes are searched, with various retrieval strategies tailored to different types of queries and datasets.

- **Similarity Search Retriever**: Often paired with vector indexes, similarity search retrievers identify documents whose vector representations closely match the query vector. This type of retriever excels at semantic search, where conceptually similar content is prioritized over exact keyword matches, and is commonly used in RAG systems, conversational agents, and QA applications.

- **Sparsity-Based Retriever**: This retriever relies on exact keyword matching and is typically used with keyword indexes. By leveraging TF-IDF or BM25, sparsity-based retrievers prioritize documents containing specific terms, making them ideal for applications that focus on term-specific searches, such as document or product searches.
- **Hybrid Retriever**: Combining the strengths of vector and sparse retrieval methods, hybrid retrievers allow for a more flexible search experience by capturing both conceptual similarity and exact keyword matches. This versatility is valuable for complex applications where both semantic relevance and keyword accuracy are important.
- **Memory-Based Retriever**: Used primarily in conversational applications, memory-based retrievers retain the context of previous interactions, enabling continuous dialogue. This continuity is essential in applications that require long-term engagement, such as customer service chatbots and virtual assistants.

LangChain's extensive suite of retrievers provides flexible options for retrieving documents, data, and context from a wide variety of sources. Each retriever is optimized for specific types of data, ensuring that users can select a solution tailored to their needs, whether for research, enterprise knowledge management, or specialized application domains. Here is more detail on some of the popular retrievers in LangChain:

- **AmazonKnowledgeBasesRetriever**: This retriever interfaces with Amazon's knowledge bases, making it suitable for enterprise environments with a vast knowledge repository in AWS. It enables streamlined access to structured corporate data and FAQs.
- **AzureAISearchRetriever**: Powered by Microsoft Azure's AI Search, this retriever offers advanced capabilities for searching through large datasets hosted on Azure. It is especially effective in enterprise settings that rely on the Azure ecosystem for data storage.
- **ElasticsearchRetriever**: Integrating directly with Elasticsearch, this retriever is highly efficient for indexing and retrieving documents based on keywords and relevance scoring, ideal for scalable search applications in both public and private databases.

- **MilvusCollectionHybridSearchRetriever**: This retriever combines vector-based and scalar searches through Milvus, an open source vector database. It is optimal for applications requiring both semantic and traditional keyword matching.
- **VertexAISearchRetriever**: Utilizing Google's Vertex AI, this retriever allows developers to perform high-quality searches across datasets managed within Google Cloud, offering seamless integration with other Google services and tools.
- **ArxivRetriever**: This retriever accesses scholarly papers directly from arXiv.org, making it perfect for academic research, literature reviews, and scientific inquiry.
- **TavilySearchAPIRetriever**: Designed for Internet-wide searches, this retriever leverages the Tavily API to bring back relevant web results, useful for general web-based information retrieval.
- **WikipediaRetriever**: Accesses content from Wikipedia, allowing users to retrieve well-organized information on a wide range of topics. It's ideal for summarizing general knowledge and historical information.
- **BM25Retriever**: BM25 is a classic algorithm in information retrieval, and this retriever brings it to LangChain without needing an external search platform. It is useful for applications requiring local, traditional keyword-based retrieval.
- **SelfQueryRetriever**: Unique in its capability, this retriever processes and interprets its own queries, offering high flexibility in search tasks where query understanding is essential.
- **MergerRetriever**: This retriever combines results from multiple retrievers, aggregating various sources to improve recall and coverage. It is highly suitable for applications needing diverse data retrieval.
- **DeepLakeRetriever**: With Deep Lake's multimodal database, this retriever accesses complex datasets, including structured and unstructured data, useful for multimedia or cross-domain projects.

- **AstraDBRetriever**: Leveraging DataStax Astra, this retriever is ideal for organizations using Cassandra-based databases, combining scalability with vector capabilities for advanced search functionality.
- **ActiveloopDeepMemoryRetriever**: By utilizing Activeloop's Deep Memory system, this retriever can store and retrieve data efficiently, making it a valuable option for high-performance applications needing rapid access to historical data.
- **AmazonKendraRetriever**: This retriever integrates with Amazon Kendra, Amazon's intelligent search service, allowing for precise and context-aware search, particularly useful in enterprise environments.
- **ArceeRetriever**: Designed for specialized and secure NLP applications, this retriever can be adapted to smaller, purpose-specific language models and secure environments.
- **BreebsRetriever**: As a retriever specifically created for the Breebs system, it provides an efficient, targeted search for users leveraging Breebs for NLP tasks.
- **AzureCognitiveSearchRetriever**: This retriever works with Azure Cognitive Search, which is well-suited for organizations in the Microsoft ecosystem, offering customizable search options and robust scalability.
- **BedrockRetriever**: By integrating with Amazon Bedrock, this retriever provides seamless retrieval capabilities within Amazon's AI suite, suitable for AWS-centric machine learning applications.

End-to-End Workflow: From Indexing to Retrieval

In LangChain's workflow, indexes and retrievers interact in a streamlined sequence:

- **Indexing Process**: Initially, the document corpus is loaded, divided into chunks, embedded, and stored within a vector or keyword index. This indexing step captures each chunk's semantic meaning and stores it in a database, like a vector database, which can be used for quick similarity searches.

- **Retrieval Process**: When a query is made, the retriever communicates with the relevant index to retrieve the most relevant chunks. Whether using similarity search, keyword matching, or both, the retriever pulls information aligned with the query. This retrieved content is then passed to the LLM, which generates a response by reasoning over the retrieved data.

Real-World Applications of LangChain Indexes and Retrievers

Retrieval-augmented generation (RAG) applications utilize LangChain's retrievers to bring in external information in real time, effectively augmenting a model's knowledge with up-to-date data. In the RAG framework, the retriever's role is to select documents relevant to a user's query, enabling the LLM to reference specific information during response generation. This process significantly enhances the quality and accuracy of generated answers by grounding them in reliable, external sources.

For instance, when the model needs to answer a question about recent scientific findings or news, RAG ensures that the most relevant and current information is retrieved and considered. This dynamic integration of external knowledge is particularly valuable in domains where accuracy and context are crucial, such as medical research, financial analysis, and technical support, as it allows the model to produce responses informed by the latest data.

In **question-answering (QA) and search systems**, LangChain's retrieval mechanisms are employed to sift through large datasets, pinpointing the specific information needed to answer direct queries. This capability is indispensable in customer support, where users frequently seek answers to targeted questions about products, services, or policies. Similarly, educational platforms leverage QA systems to help students and researchers retrieve information from vast databases or digital libraries.

Here, the retriever works by filtering through indexed content and extracting the passages most relevant to the query. By presenting the most pertinent information first, QA applications powered by LangChain's retrievers improve user satisfaction, reduce search time, and increase the precision of answers. Research assistants and document-heavy industries, such as law and academia, can also benefit from QA systems, as they streamline the retrieval of highly specific knowledge from expansive collections of information.

Conversational agents leverage LangChain's memory-based retrievers to create engaging and personalized dialogue experiences. Unlike typical retrieval tasks, conversational applications require continuity, as users expect the system to "remember" prior exchanges and respond contextually. Memory-based retrievers enable these systems to track and recall relevant information from previous interactions, allowing the agent to build upon past conversations. This is particularly advantageous in customer service, where understanding a user's past queries can help address ongoing issues more effectively, and in personal assistant applications, where maintaining familiarity with a user's preferences and history enhances personalization.

For example, in virtual health assistants, memory retention enables the agent to remember past symptoms or medical advice, providing users with a consistent and coherent experience across multiple interactions.

Overall, LangChain's indexes and retrievers are foundational in building robust, adaptable applications that deliver real-time, accurate, and contextually aware responses. From dynamically pulling the latest information for RAG to improving efficiency in QA systems and fostering continuity in conversational agents, these tools support a wide range of real-world use cases that require precise, responsive, and intelligent information retrieval.

Example:

```
from langchain.document_loaders import TextLoader
from langchain.embeddings import OpenAIEmbeddings
from langchain.vectorstores import FAISS
from langchain.chat_models import ChatOpenAI
from langchain.chains import RetrievalQA
from langchain.docstore import InMemoryDocstore
from langchain.schema import Document
import faiss

# Step 1: Prepare Sample Documents
documents = [
    Document(page_content="Italy is a beautiful country in Europe, known
    for its rich history and culture. It has famous landmarks like the
    Colosseum and Leaning Tower of Pisa.", metadata={"title": "About
    Italy"}),
```

```
    Document(page_content="Italian cuisine is popular worldwide, with
    dishes like pasta, pizza, and gelato. Each region in Italy has its own
    unique culinary specialties.", metadata={"title": "Italian Cuisine"}),
    Document(page_content="Rome is the capital city of Italy, known for its
    ancient history and architecture, including the Vatican City and the
    Pantheon.", metadata={"title": "Rome - The Capital"})
]

# Step 2: Initialize Embeddings and Vector Store
embedding_model = OpenAIEmbeddings()
embedding_dim = 1536  # Ensure this matches the embedding model
index = faiss.IndexFlatL2(embedding_dim)

# Set up FAISS vector store
vector_store = FAISS(
    embedding_function=embedding_model.embed_query,
    index=index,
    docstore=InMemoryDocstore({}),  # Empty docstore to start
    index_to_docstore_id={}          # Start with an empty mapping
)

# Add documents to the vector store
vector_store.add_documents(documents)

# Step 3: Set Up the Retrieval-Enhanced Generation (RAG) Chain
retriever = vector_store.as_retriever()
llm = ChatOpenAI(model="gpt-3.5-turbo")

# Create the RAG chain
rag_chain = RetrievalQA.from_chain_type(
    llm=llm,
    chain_type="stuff",
    retriever=retriever
)

# Step 4: Ask Questions and Get Answers
questions = [
    "Tell me about Italy.",
```

```
    "What food is Italy famous for?",
    "What are some historical sites in Rome?"
]

for question in questions:
    answer = rag_chain.run(question)
    print(f"Question: {question}")
    print(f"Answer: {answer}\n")
```

Output:

```
Question: Tell me about Italy.
Answer: Italy is a beautiful country in Europe, known for its rich history,
culture, and stunning landscapes. It has famous landmarks like the
Colosseum in Rome and the Leaning Tower of Pisa. Rome is the capital city
of Italy, famous for its ancient history and architecture, including the
Vatican City and the Pantheon. Italian cuisine is popular worldwide, with
dishes like pasta, pizza, and gelato being well-loved. Each region in Italy
has its own unique culinary specialties, making it a food lover's paradise.
Question: What food is Italy famous for?
Answer: Italy is famous for dishes like pasta, pizza, gelato, risotto, and
tiramisu. Each region in Italy has its own unique culinary specialties,
making Italian cuisine diverse and beloved worldwide.
Question: What are some historical sites in Rome?
Answer: Some historical sites in Rome include the Colosseum, Roman Forum,
Pantheon, and the Vatican City.
```

Using LangChain Indexing API

In this chapter, we explore a foundational workflow for indexing documents using LangChain's indexing API. This API allows you to import and synchronize documents from various sources into a vector store, offering a systematic approach to managing document data for efficient retrieval. The indexing API supports a range of optimizations, ensuring that documents are indexed only when necessary, thus saving both time and computational resources.

One of the primary advantages of the indexing API is its ability to prevent redundant content in the vector store. By avoiding the reindexing of unchanged documents and skipping duplicate content, this tool significantly enhances retrieval speed and efficiency. Furthermore, the API avoids recalculating embeddings for previously indexed documents unless they have been altered, ensuring that the vector store is always up-to-date without wasting resources. This workflow aligns particularly well with vector search applications, where precision and efficiency in document retrieval are paramount.

Technical Structure of the Indexing API

LangChain's indexing process relies on a robust mechanism managed by a component called the **RecordManager**. This manager serves as a tracker, logging each document addition to the vector store with essential metadata.

Each document receives a unique hash—a digital signature that represents the content of both the text and its metadata. This hash, alongside the time of writing and the document's source identifier, enables the system to maintain efficient, organized indexing, even when documents undergo several stages of transformation, such as text chunking, which divides lengthy texts into smaller, manageable sections for indexing.

Deletion Modes and Content Maintenance

To maintain an efficient vector store, LangChain offers several deletion modes to handle outdated or redundant documents. Three main modes—**None**, **Incremental**, and **Full**—each provide different levels of automation for clearing old or modified data.

- The **None** mode requires manual cleanup, allowing developers to directly manage obsolete content.
- **Incremental** mode continuously clears out old data as it processes new content, efficiently minimizing outdated entries.
- **Full** mode, in contrast, performs a complete cleanup after each batch of documents is indexed, ensuring that no old or duplicate data remains.

For example, if the content of a document changes, both **Incremental** and **Full** modes will delete the previous version from the vector store. However, if a source document is removed entirely, **Full** mode will erase it automatically, while **Incremental** will not. This staged approach to deletion ensures the accuracy and efficiency of the indexing process while maintaining data integrity.

In cases where documents are modified, there may be a brief interval in which both the old and new versions coexist in the store. **Incremental** mode minimizes this overlap, as it cleans up continuously. **Full** mode, however, clears outdated data only after all new data has been processed, which may lead to a slightly longer overlap period.

Requirements and Compatibility

For optimal functionality, it's recommended to use LangChain's indexing API with vector stores that support document management by ID, as this enables precise addition and deletion operations. Notably, the indexing API is compatible with a wide range of vector stores, including popular options like Pinecone, Redis, FAISS, and Weaviate. Each of these stores supports key features such as `add_documents` and `delete` methods with ID arguments, which allow for accurate document management.

Compatible Vector Stores:

- **Aerospike**: High-performance, scalable database for real-time data storage and retrieval
- **AstraDB**: Distributed, cloud-native database built on Apache Cassandra for scalable applications
- **Azure Cosmos DB NoSQL/Vector Search**: Microsoft's scalable NoSQL database with vector search capabilities
- **Cassandra**: Open source, distributed database designed for scalability and reliability
- **Chroma**: Vector database optimized for handling high-dimensional data
- **Databricks Vector Search**: Integrated vector search within Databricks for enhanced data processing
- **DeepLake**: Vector database for machine learning datasets, optimized for deep learning workflows

- **Elastic Vector Search**: Vector-based search support within Elasticsearch for relevant data insights
- **FAISS**: Open source library for fast, approximate nearest neighbor search, commonly used for vector search
- **Milvus**: Open source, cloud-native vector database optimized for high-performance similarity search
- **MongoDB Atlas Vector Search**: MongoDB's vector search capabilities for enhanced data retrieval
- **Pinecone**: Fully managed vector database for real-time search and machine learning applications
- **Qdrant**: High-performance vector database supporting semantic search and similarity matching
- **Redis**: In-memory datastore with modules supporting vector search for low-latency applications
- **SingleStoreDB**: Distributed SQL database optimized for real-time analytics and vector search
- **Supabase Vector Store**: Open source alternative to Firebase with vector storage capabilities
- **Vespa Store**: Open source platform for real-time indexing and serving of large datasets
- **Weaviate**: Open source vector search engine with built-in NLP support for semantic search
- **Tencent VectorDB**: Vector database by Tencent for fast, efficient data retrieval in AI applications

Important Considerations

LangChain's RecordManager uses a timestamp-based mechanism for determining when content should be cleaned. However, in rare situations where two tasks execute consecutively within a very short time interval, this mechanism may experience limitations, potentially leaving some content temporarily unprocessed. This issue is unlikely in practical applications, as the RecordManager uses high-resolution timestamps, and indexing tasks typically take more than a few milliseconds to complete. This time-based approach helps ensure accuracy while preserving system performance and responsiveness.

Agents in LangChain

In the evolving landscape of artificial intelligence, language models (LLMs) and frameworks like LangChain have redefined our approaches to data analysis, information synthesis, and content generation. At the heart of these capabilities lies the concept of *agents*—intelligent systems that employ LLMs to orchestrate complex tasks and make informed decisions. In this chapter, we'll delve into the dual roles agents play in harnessing LLMs: as content generators and as reasoning engines.

Leveraging their extensive pretrained knowledge, LLMs can function as content generators, producing unique, engaging content from scratch. Alternatively, when deployed as reasoning engines, they synthesize and manage information from multiple sources, analyzing data and planning actionable steps. Both approaches bring distinct advantages and challenges, with the optimal use case determined by the task's specific needs.

Defining Agents

In the context of LLMs, *agents* facilitate the decision-making process by determining what actions to take and in what sequence. These actions can include using a tool, observing the results, or generating a response for the user. Effective use of agents allows AI systems to operate with precision and adaptability.

Agents in LangChain, for instance, employ a high-level API to streamline complex interactions and decision-making processes. Before diving into practical applications, understanding key terms is essential:

- **Tool**: A designated function for performing a specific task, such as conducting a Google Search, querying a database, or executing code in a Python environment. A tool's interface typically consists of a function that accepts a string input and returns a string output.
- **Language Model (LLM)**: The core language model that powers the agent, responsible for understanding and generating text.

- **Agent**: The orchestrating system that integrates LLMs and tools, executing commands based on user input and contextual cues. LangChain supports several standard agents accessible through the high-level API, and customized agents can also be implemented as needed.

Types of Agents in LangChain

Currently, most agents in LangChain fall into two primary categories:

1. **Action Agents**: Designed for direct, single-action tasks, Action Agents execute straightforward commands and are ideal for brief, specific interactions.
2. **Plan-and-Execute Agents**: These agents take a broader approach, planning a sequence of actions to achieve a goal and executing each action step-by-step. This type is suited for complex, long-term tasks that require sustained focus. However, the extended planning process may result in increased latency. A practical approach is to employ an Action Agent within a Plan-and-Execute Agent's workflow, allowing for efficiency without sacrificing depth.

In a typical Action Agent workflow

1. The agent receives user input and selects the appropriate tool or action.
2. The chosen tool is activated, and its output (or "observation") is recorded.
3. The observation, along with the history of actions, is passed back to the agent to guide the next step.
4. The agent iterates through this process until it determines no further actions are required, at which point it provides a direct response to the user.

Tools As Extensions of Language Models

Agents gain flexibility and relevance through the use of *tools*, which extend the capabilities of LLMs by interfacing with external data sources, APIs, and computational resources. Tools enable agents to access up-to-date information, run code, and interact with files—crucial functions given that LLMs are often limited to static, pretrained data. By incorporating tools, agents can enrich the LLM's understanding with real-time data and more precise context, thereby enhancing its decision-making ability.

Content Generation vs. Reasoning Engines

When employing an LLM through agents, two primary modes of operation emerge: content generation and reasoning.

1. **Content Generators**: In this role, an LLM produces content purely from its internal knowledge, drawing upon a rich reservoir of pretrained data to create unique and creative outputs. However, this can also result in unverified or speculative information, often referred to as "hallucinations."

2. **Reasoning Engines**: When acting as a reasoning engine, the agent functions more as an information manager than a creator. In this mode, it seeks to gather, verify, and synthesize relevant information, frequently with the aid of external tools. The LLM draws on data sources related to the topic and constructs new, accurate content by summarizing and integrating critical insights.

By understanding these dual modes—content generation and reasoning—users can better tailor LLM-powered agents to meet diverse task requirements, from creative writing to intricate data analysis, thus maximizing the model's potential in each unique application.

Example:

```
from langchain.chat_models import ChatOpenAI
from langchain.embeddings import OpenAIEmbeddings
from langchain.vectorstores import FAISS
from langchain.agents import initialize_agent, AgentType
from langchain.tools import Tool
```

```
from langchain.text_splitter import CharacterTextSplitter
from langchain.docstore.document import Document
import os

# Define document contents
document1_content = "The capital of France is Paris. Paris is known for its
art, fashion, and culture."
document2_content = "The capital of Japan is Tokyo. Tokyo is famous for its
technology and vibrant city life."

# Create Document objects
documents = [
    Document(page_content=document1_content),
    Document(page_content=document2_content)
]

# Split text into chunks for vector indexing
text_splitter = CharacterTextSplitter(chunk_size=1000, chunk_overlap=100)
docs = text_splitter.split_documents(documents)

# Create an embedding model for indexing
embedding = OpenAIEmbeddings()

# Create a FAISS vector store with the documents and embeddings
vector_store = FAISS.from_documents(docs, embedding)

# Initialize OpenAI model using Chat API with gpt-3.5-turbo
llm = ChatOpenAI(model="gpt-3.5-turbo")

# Define a tool to query the vector store
def query_vector_store(query: str) -> str:
    results = vector_store.similarity_search(query, k=1)
    return results[0].page_content if results else "No relevant
    information found."

tools = [
    Tool(
        name="Document Index",
        func=query_vector_store,
```

```
        description="Use this tool to answer questions about the capital
        cities in the documents."
    )
]

# Set up the agent
agent = initialize_agent(
    tools=tools,
    llm=llm,
    agent=AgentType.ZERO_SHOT_REACT_DESCRIPTION,
    verbose=True
)

# Ask a question
question = "What is the capital of Japan?"
response = agent({"input": question})
print(response["output"])
```

How it works:

- **The agent receives a natural language query ("What is the capital of Japan?").**
- It analyzes the input and sees that there's a tool available called "Document Index" with a description suggesting it's useful for capital city questions.
- Using the **ReAct reasoning**, it decides to call the tool (query_vector_store(...)) with the input.
- The tool queries the FAISS vector store for relevant info and returns the most relevant chunk.
- The agent then returns the final answer.

AgentType.ZERO_SHOT_REACT_DESCRIPTION ells LangChain to use an agent that:

- Uses reasoning and tools in a step-by-step fashion (ReAct),

- Figures everything out without seeing examples (zero-shot),
- Relies on tool **names and descriptions** to choose the right action.

Output:

```
> Entering new AgentExecutor chain...
I should use the Document Index tool to search for the capital of Japan in
the documents.
Action: Document Index
Action Input: "capital of Japan"
Observation: The capital of Japan is Tokyo. Tokyo is famous for its
technology and vibrant city life.
Thought:I now know the final answer
Final Answer: The capital of Japan is Tokyo.

> Finished chain.
The capital of Japan is Tokyo.
```

Exploring Autonomous Agents: AutoGPT and BabyAGI

AutoGPT and BabyAGI represent groundbreaking advancements in the realm of autonomous agents—AI systems designed to accomplish tasks with minimal human supervision. Their unique ability to independently work toward specific objectives has garnered significant attention, with AutoGPT amassing over 100,000 stars on GitHub and sparking global curiosity. These agents offer a glimpse into the future of AI-driven autonomy and promise transformative applications across various domains.

AutoGPT, an open source platform, utilizes GPT-4 to systematically explore the Internet, decompose complex tasks into manageable subtasks, and even initiate new agents to help achieve its goals. BabyAGI operates similarly, integrating GPT-4, a vector store, and LangChain to create tasks based on prior outcomes and a primary objective. Although still in development, both systems highlight the immense potential of autonomous agents and underscore their rapid progress and broad applicability.

Autonomous agents like AutoGPT and BabyAGI appeal to the AI community for three main reasons:

- **Minimal Human Involvement**: Unlike traditional models that rely on human input (e.g., ChatGPT), these agents require little guidance to operate.
- **Diverse Applications**: From personal assistance to task automation, their potential use cases are expansive.
- **Rapid Development**: The swift evolution of these technologies signals their potential to revolutionize various industries.

To optimize the performance of autonomous agents, it is essential to set well-defined goals, which might include generating natural language content, providing accurate responses, or refining actions based on user feedback.

What Is AutoGPT?

AutoGPT is an autonomous agent capable of operating independently until it reaches a specified goal. This agent leverages three core features:

1. **Internet Connectivity**: AutoGPT accesses the web in real time, allowing for ongoing research and information gathering.
2. **Self-Prompting**: It generates and organizes subtasks autonomously to tackle larger goals.
3. **Task Execution**: AutoGPT executes tasks, including activating additional AI agents. However, this feature sometimes encounters challenges, such as task loops or misinterpretations.

Initially conceived as a general-purpose agent capable of handling virtually any task, AutoGPT's broad scope revealed limitations in efficiency. Consequently, the trend has shifted toward developing specialized agents tailored for specific tasks, thus enhancing their practical utility.

How Does AutoGPT Work?

AutoGPT's design allows it to go beyond simple text generation, transforming it into a task-oriented agent capable of creating, prioritizing, and executing complex sequences of actions. This operational model allows AutoGPT to

1. Understand overarching goals
2. Break goals into subtasks
3. Execute tasks
4. Adjust actions based on contextual information

AutoGPT relies on plug-ins for Internet browsing and other external access. Its memory module stores context, enabling it to evaluate situations, self-correct, and reprioritize as necessary. This dynamic feedback loop allows AutoGPT to perform as a proactive, goal-oriented agent rather than a passive language model.

This independence opens new possibilities in AI-driven productivity but introduces challenges around control, unintended consequences, and ethical considerations.

What Is BabyAGI?

Like AutoGPT, BabyAGI is an autonomous agent designed to operate continuously, drawing from a task list, executing actions, and creating new tasks based on previous outcomes. However, BabyAGI employs a distinct approach, integrating four specialized sub-agents to manage its operations:

1. **Execution Agent**: Executes tasks by constructing prompts based on the objective and feeding them to a language model (e.g., GPT-4)
2. **Task Creation Agent**: Generates new tasks from prior task results and objectives, creating a list of new tasks
3. **Prioritization Agent**: Orders tasks based on urgency or importance
4. **Context Agent**: Merges results from previous executions to maintain continuity across tasks

Key Features of BabyAGI

BabyAGI exemplifies the potential for autonomous agents to manage and adapt to complex workflows:

- **Autonomous Task Management**: BabyAGI dynamically generates new tasks and reprioritizes its task list in response to updated goals or information.
- **Efficient Storage and Search**: BabyAGI uses GPT-4 for task execution, a vector database for efficient data storage, and LangChain for decision-making.
- **Adaptability**: BabyAGI not only completes tasks but also enriches and stores results in a database, enabling it to learn and evolve based on new data.

This integration of GPT-4 and LangChain capabilities allows BabyAGI to interact with its environment and perform efficiently within defined constraints.

A Practical Implementation of BabyAGI

BabyAGI's implementation with LangChain provides flexibility; while it currently uses a FAISS vector store, users can adapt it to other storage solutions. In a recent update (as of August 2023), LangChain reorganized some experimental features, moving them to a new library called `langchain_experimental`. To implement BabyAGI with the updated LangChain library, install the experimental package and modify code references accordingly.

AutoGPT and BabyAGI offer a fascinating look into the potential of autonomous AI. Through continuous innovation, these agents are setting the stage for future AI systems capable of independent operation, complex decision-making, and task execution across diverse environments. Whether streamlining workflows, managing data, or providing real-time assistance, autonomous agents promise to transform AI from a reactive tool into a proactive, learning system poised to reshape the boundaries of human–AI collaboration.

LLM Models in LangChain

Chat Models

AI21 Labs

AI21 Labs offers models designed for natural language understanding and generation, tailored to enhance interaction in various applications.

Alibaba Cloud PAI EAS

A lightweight, cost-effective AI solution from Alibaba Cloud, PAI EAS facilitates scalable deployments and high-performance machine learning, suitable for a range of business applications and data-driven insights.

Anthropic

Anthropic's conversational models prioritize safe and interpretable AI interactions, offering reliable tools and guidance for integration into projects requiring advanced language understanding.

Anyscale

Anyscale's integration with LangChain allows seamless access to scalable chat models, suitable for enhancing complex AI applications and workflows in diverse environments.

Azure OpenAI

Microsoft Azure's OpenAI integration enables developers to deploy and scale OpenAI's advanced language models, optimized for a range of applications from customer service to sophisticated content generation.

Azure ML Endpoint

A comprehensive platform by Azure for building, training, and deploying machine learning models, Azure ML Endpoint allows for streamlined deployment of conversational AI with enterprise-grade scalability.

Baidu Qianfan

A unified platform from Baidu AI Cloud, Qianfan offers end-to-end solutions for large model development, from training and deployment to performance tuning and scaling.

AWS Bedrock

Amazon's AWS Bedrock provides a foundation for deploying conversational AI models at scale, backed by robust infrastructure for handling various use cases, including customer service, virtual assistants, and more.

Cohere

With a focus on accessible language models, Cohere provides tools for natural language understanding, enabling quick deployment of conversational AI into customer-facing or internal applications.

Databricks

The Databricks Lakehouse Platform unifies data, analytics, and AI, providing an integrated solution that allows organizations to develop, train, and deploy chat models efficiently on a single platform.

DeepInfra

DeepInfra's serverless AI inference service provides easy access to conversational models, offering a cost-effective, scalable way to deploy natural language applications without extensive infrastructure.

Eden AI

Eden AI aggregates top-tier AI models, uniting various providers under one platform, enabling developers to seamlessly integrate and compare multiple chat solutions in their applications.

EverlyAI

EverlyAI allows users to scale machine learning models in the cloud, providing robust solutions for integrating conversational AI into applications that need to handle high-volume interactions.

Fireworks

Fireworks AI chat models offer powerful language capabilities tailored for customer service, education, and content generation, designed to help businesses implement responsive, intuitive AI.

GigaChat

Integrated with LangChain, GigaChat enables the development of conversational AI with a focus on providing adaptive, context-sensitive responses, ideal for interactive applications.

Google AI

Google AI offers a comprehensive suite of chat models designed for seamless interaction, optimized to support complex, multiturn conversations in various application scenarios.

Google Cloud Vertex AI

Vertex AI on Google Cloud delivers advanced chat model solutions, allowing developers to train, optimize, and deploy large language models that drive enhanced user interactions.

GPTRouter

GPTRouter serves as an open source API gateway, enabling easy access and routing across various large language models, simplifying the deployment of conversational AI in diverse projects.

Groq

Groq's chat models provide a high-speed solution for conversational AI, helping businesses implement responsive, scalable models that perform well in interactive, real-time environments.

ChatHuggingFace

Hugging Face offers an extensive library of chat models that can be easily integrated with LangChain, allowing developers to experiment with and deploy a variety of conversational AI solutions.

IBM watsonx.ai

IBM's watsonx.ai foundation models are designed for enterprise-grade conversational AI, providing reliable and secure solutions for handling complex customer interactions and data management.

JinaChat

JinaChat's models bring efficient natural language processing capabilities to a range of applications, making it easy to integrate responsive AI into both customer-facing and internal platforms.

Kinetica

Kinetica's AI tools support transforming natural language into actionable data insights, making it a valuable platform for conversational AI that interacts with and analyzes real-time data.

LiteLLM

LiteLLM provides simplified access to major language models like Anthropic, Azure, and Hugging Face, streamlining the deployment of conversational AI in diverse applications.

LiteLLM Router

LiteLLM's Router enables seamless integration and routing among various chat model providers, offering flexibility and ease of management across different AI platforms.

Llama 2 Chat

Llama 2 Chat integrates Llama-2 large language models with additional chat capabilities, creating a robust tool for applications requiring natural language understanding and conversation.

Llama API

LlamaAPI offers hosted language models through LangChain, allowing developers to deploy and manage interactive conversational applications efficiently.

LlamaEdge

LlamaEdge enables local and cloud-based deployment of LLMs in GGUF format, providing flexible, efficient options for integrating chat capabilities.

Llama.cpp

The Llama.cpp Python library provides simple bindings for lightweight Llama models, making it easier to integrate and experiment with conversational AI solutions.

maritalk

Maritalk introduces its conversational models with a focus on responsive, user-friendly dialogue capabilities suitable for various customer-facing applications.

MiniMax

MiniMax offers large language models geared toward enterprise applications, providing reliable, scalable solutions for complex conversational tasks.

MistralAI

MistralAI offers robust tools and guidance for deploying conversational models that can handle multiturn interactions in diverse applications.

MLX

MLX's chat models facilitate conversational AI use, helping developers integrate intuitive and responsive dialogue systems into their projects.

Moonshot

Moonshot, a Chinese startup, provides enterprise-focused large language models, offering scalable AI solutions for businesses across various industries.

Naver

Naver provides an intuitive platform for conversational AI, enabling users to deploy and customize chat models for interactive applications.

NVIDIA AI Endpoints

NVIDIA's AI Endpoints deliver high-performance chat models that cater to complex interactions, making it suitable for applications requiring advanced conversational capabilities.

ChatOCIModelDeployment

Oracle's OCIModelDeployment chat models offer seamless integration within Oracle's ecosystem, facilitating enterprise-grade conversational AI.

OCIGenAI

Oracle's GenAI models allow users to leverage AI capabilities with a focus on reliability, scalability, and seamless deployment across diverse environments.

ChatOctoAI

OctoAI provides access to efficient compute resources, enabling developers to integrate fast and responsive conversational AI models into their projects.

Ollama

Ollama allows users to run open source models like LLaMA either locally or in the cloud, providing a flexible solution for deploying conversational AI.

OpenAI

OpenAI's chat models offer advanced language capabilities, making it easy to implement responsive and accurate conversational AI for various use cases.

Perplexity

Perplexity AI models offer tools for natural language processing and conversational AI, supporting accurate and dynamic user interactions.

PremAI

PremAI is an all-in-one platform simplifying the development of robust chat applications, helping users create, train, and deploy conversational AI quickly.

PromptLayer ChatOpenAI

PromptLayer connects with OpenAI models to log and track interactions, making it easier to monitor and improve conversational performance.

SambaNovaCloud

SambaNovaCloud's chat models provide scalable conversational AI options, ideal for applications with high volumes of complex interactions.

SambaStudio

SambaStudio facilitates the deployment and management of chat models, offering a comprehensive solution for organizations looking to integrate conversational AI.

Snowflake Cortex

Snowflake Cortex integrates large language models directly within the Snowflake platform, enabling seamless access to chat models alongside analytics.

solar

Solar-powered AI solutions for sustainable applications in natural language processing and conversational AI.

SparkLLM Chat

iFlyTek's SparkLLM offers a powerful conversational AI model, providing high-quality language understanding and interaction capabilities.

Nebula (Symbl.ai)

Nebula from Symbl.ai specializes in conversation analytics, supporting businesses with models designed for complex conversational analysis and interaction.

Tongyi Qwen

Alibaba's DAMO Academy developed Tongyi Qwen, a large language model offering advanced conversational capabilities suitable for various applications.

Upstage

Upstage's chat models are designed for quick integration, providing a flexible solution for conversational AI in customer service and engagement applications.

vLLM Chat

vLLM can be deployed to mimic the OpenAI API, offering users a flexible, compatible chat model that integrates easily into existing workflows.

Volc Enging Maas

Volc Enging Maas chat models offer a scalable AI platform for businesses looking to integrate conversational AI solutions.

YandexGPT

YandexGPT's models are available via LangChain, enabling integration with Yandex's conversational AI for localized and global applications.

Supported LLMs

- **AI21 Labs**: Juristic models for legal and technical language in natural interactions
- **Aleph Alpha**: Luminous models for text understanding and generation, ideal for content creation and support
- **Alibaba Cloud PAI EAS**: Comprehensive platform for scalable AI model training, deployment, and management

- **Amazon API Gateway**: Managed API service for easy deployment and management of back-end services
- **Anyscale**: Fully managed Ray platform for distributed AI applications
- **Azure ML**: End-to-end platform for building, training, and deploying machine learning models
- **Azure OpenAI**: Deployment of OpenAI models through Azure for advanced NLP
- **Baichuan LLM**: Large language model focused on conversational AI for health and well-being
- **Baidu Qianfan**: Platform for training, deploying, and optimizing large language models on Baidu AI Cloud
- **Baseten**: Simplifies model deployment and operation within the LangChain ecosystem
- **Beam**: API wrapper for deploying large language models with scalable resources
- **Bedrock (Amazon)**: Documentation for integrating NLP models within Amazon's infrastructure
- **Clarifai**: AI platform for managing the full AI life cycle, from data preparation to deployment
- **Cloudflare Workers AI**: Edge-deployed generative models for low-latency language AI
- **Cohere**: Models for natural language processing, enhancing language understanding and interaction
- **Databricks**: Lakehouse platform for unified data, analytics, and AI model management
- **DeepInfra**: Serverless AI service for easy deployment of language models
- **Eden AI**: Aggregated API access to top AI models, supporting diverse AI integrations

- **ExLlamaV2**: Optimized library for running large models on local hardware
- **ForefrontAI**: Platform for fine-tuning and deploying open source language models
- **GigaChat**: Tools for interactive conversational AI in dynamic environments via LangChain
- **Google Vertex AI**: Model deployment and scaling tools for machine learning workflows
- **GPT4All**: Open source ecosystem for robust conversational agents
- **Gradient**: Supports model fine-tuning and deployment, integrated with LangChain
- **Hugging Face**: Extensive model repository for deploying and managing NLP models
- **IBM watsonx.ai**: Enterprise-grade tools for managing large language models
- **Intel IPEX-LLM**: PyTorch library optimized for running models on Intel CPUs/GPUs
- **Llama.cpp**: Lightweight bindings for Llama models in Python applications
- **Minimax**: Chinese startup providing NLP services and conversational AI
- **Modal**: Serverless compute platform for easy AI deployment
- **MosaicML**: Managed inference for NLP applications, supporting model customization
- **NLP Cloud**: Scalable NLP models for companies via API
- **NVIDIA**: High-performance model deployment on NVIDIA hardware
- **Oracle Generative AI**: Oracle's scalable infrastructure for AI model training and deployment

- **OpenAI**: Guidance for integrating OpenAI's models in various applications
- **OpenLLM**: Open platform compatible with OpenAI's API for model management
- **OpenVINO**: Toolkit for running AI models on Intel hardware
- **Replicate**: Cloud platform for easy access and deployment of AI models
- **SageMaker**: Amazon's platform for building and deploying machine learning models
- **SambaNova**: Tools for running and managing open source models in enterprise applications
- **SparkLLM**: iFlyTek's large language model for complex NLP tasks
- **StochasticAI**: Platform for AI model life cycle management
- **TextGen**: Gradio-based web UI for interactive content generation
- **Titan Takeoff**: Tools for small, efficient language models in business
- **Together AI**: Collaborative language models for distributed environments
- **Tongyi Qwen**: Alibaba's model for broad NLP applications
- **Writer**: AI-driven platform for generating multilingual content for marketing and writing
- **Xorbits Inference (Xinference)**: Scalable library for large language model serving
- **YandexGPT**: Integration support for multilingual capabilities with YandexGPT

LLMs vs. Chat Models

Large Language Models (LLMs)

Large language models are AI systems trained on vast amounts of text data to perform a broad range of language tasks, such as summarization, text generation, translation, and sentiment analysis. LLMs, like OpenAI's GPT, are primarily designed for general-purpose language processing and can be adapted to various applications by using prompt engineering, fine-tuning, or transfer learning. They generate responses based on context without specific training for conversational flows, making them versatile but less specialized for natural, interactive dialogue.

Key Characteristics of LLMs

- **General Purpose**: Capable of handling a broad spectrum of language tasks
- **Few-Shot and Zero-Shot Learning**: Can handle tasks with minimal examples or prompts
- **Less Interactive**: Not optimized specifically for dynamic conversation or managing turns in a dialogue

Chat Models

Chat models are specialized derivatives or adaptations of LLMs fine-tuned specifically for conversational AI, making them better suited to applications like customer service bots, virtual assistants, or real-time chat interactions. These models have been trained on conversational data, allowing them to understand and manage conversational nuances such as tone, context retention, multiturn dialogue, and even empathy. They are optimized to handle back-and-forth interactions with users and manage context over extended exchanges.

Key Characteristics of Chat Models

Here are the key characteristics of chat models (Table 2-1):

- **Dialogue-Focused**: Trained on conversational data for a more interactive, turn-based flow
- **Context Management**: Maintains context across multiple dialogue turns, supporting natural back-and-forth interactions
- **User Alignment**: Often refined for specific use cases like customer support, virtual assistants, and real-time conversation

Table 2-1. *Key Characteristics of Chat Models*

Feature	LLMs	Chat Models
Purpose	Broad language tasks	Optimized for conversation
Training Data	General Internet or document data	Conversational data
Context Management	Limited in longer interactions	Manages context across exchanges
Interaction Style	One-off responses	Multiturn, interactive
Use Cases	Content creation, summarization, analysis, image analysis and generation	Chatbots, customer service, virtual assistants

- **AI21**: High-quality embeddings in LangChain for tasks like information retrieval, recommendations, and text similarity
- **Aleph Alpha**: Semantic embeddings with Luminous models for document comparison and search
- **Anyscale**: Embeddings optimized for distributed AI applications, supporting large-scale deployments
- **AwaDB**: AI-native database focused on scalable embedding storage and retrieval for AI insights
- **AzureOpenAI**: Scalable embedding models for intelligent search and text applications
- **Baidu Qianfan**: Unified platform for embedding and model management on Baidu AI Cloud
- **Bedrock (Amazon)**: Managed embedding service with diverse models for NLP applications

- **BGE on Hugging Face**: High-quality vector embeddings for search and retrieval
- **Clarifai**: End-to-end AI platform with embedding generation and data management tools
- **Cloudflare Workers AI**: Distributed embeddings with reduced latency for global data access
- **Cohere**: Embeddings in LangChain for natural language understanding and data-centric applications
- **Databricks**: Lakehouse platform integrating embeddings with large-scale data processing
- **DeepInfra**: Serverless embeddings for real-time applications
- **EDEN AI**: Platform with diverse embedding options for search, categorization, and similarity scoring
- **Elasticsearch**: Embedding support for enhanced search relevance and data-driven insights
- **FastEmbed by Qdrant**: Lightweight, high-speed embedding library for real-time applications
- **Fireworks**: Flexible embeddings for search and clustering in LangChain
- **GigaChat**: Efficient embeddings for AI-driven applications and high-interaction tasks
- **Google Vertex AI**: Enterprise embeddings optimized for large-scale data management
- **GPT4All**: Local embeddings focused on privacy for offline applications
- **Gradient**: Platform for embedding generation and fine-tuning for specific data needs
- **Hugging Face**: Versatile embeddings accessible through LangChain for NLP workflows

- **IBM Watsonx.ai**: Enterprise-grade semantic embeddings for data processing and analytics
- **Intel Transformers**: Optimized, quantized embeddings for efficient data representation
- **Jina**: Embedding support for search, recommendation, and indexing
- **John Snow Labs**: Healthcare-focused embeddings for scientific text
- **LASER by Meta AI**: Multilingual embeddings for cross-lingual applications
- **Llama.cpp**: Efficient bindings for Llama embeddings in constrained environments
- **LocalAI**: Local, cloud-free embeddings for secure data processing
- **MiniMax**: Robust embeddings supporting complex NLP tasks
- **ModelScope**: Repository with multilingual embedding options
- **MosaicML**: Managed embeddings for scalable, customizable data representation
- **Naver**: High-performance embeddings for search, translation, and indexing
- **NLP Cloud**: Secure, fast embeddings for reliable semantic tasks
- **NVIDIA NIMs**: NVIDIA-optimized embeddings for high-performance AI
- **Oracle Generative AI**: Scalable, managed embeddings for enterprise applications
- **OpenAI**: High-quality embeddings for similarity matching and clustering
- **OpenClip**: Open source multimodal embeddings linking text and images
- **OpenVINO**: Intel-optimized embeddings for efficient language model deployment

- **Oracle AI Vector Search**: Embeddings for AI-driven database applications
- **Pinecone**: Vectorized storage and retrieval powering search and recommendation
- **SageMaker (Amazon)**: Large-scale embedding generation for managed AI infrastructure
- **SambaNova**: Scalable embeddings for complex data needs in AI applications
- **Sentence Transformers on Hugging Face**: High-quality embeddings for search and clustering
- **SpaCy**: Embeddings for text classification and similarity tasks
- **TensorFlow Hub**: Pretrained models for NLP embedding deployment
- **TextEmbed**: REST API for scalable, low-latency embedding generation
- **Titan Takeoff**: Lightweight embeddings for cost-effective AI in business
- **Together AI**: Collaborative embeddings for distributed applications
- **Voyage AI**: Advanced embeddings for analytics and recommendations
- **YandexGPT**: Multilingual embeddings for diverse language tasks

Instruct Models

Instruct models are a specialized class of language models fine-tuned to follow natural language instructions, making them highly effective for task-oriented and interactive applications. Unlike base models—which are primarily trained to predict the next word in a sequence—instruct models are designed to interpret user input as a directive and respond with relevant, goal-focused outputs. This makes them especially useful in frameworks like LangChain, where agents are expected to make decisions, use tools, and complete multistep reasoning tasks based on a single user prompt.

In LangChain, instruct models are crucial for agent types, which require the model to understand tool descriptions, choose the right tools, and execute tasks in a step-by-step manner using the ReAct (Reasoning and Acting) paradigm. For example, when a user asks, "What is the capital of Japan?", an instruct model can identify that a knowledge lookup is needed, select a vector search tool, retrieve the relevant information, and present a concise answer—all without requiring hard-coded logic or examples.

These models work exceptionally well in scenarios where clarity, precision, and contextual relevance are important. They reduce the need for complex prompt engineering and support a wide range of applications such as question answering, summarization, data extraction, content generation, and conversational agents.

Key Benefits of Instruct Models

- Follow natural language instructions without needing detailed prompt formatting
- Understand and use external tools when paired with agent frameworks
- Perform multistep reasoning and planning
- Ideal for applications requiring structured output or task completion

A Comprehensive List of Popular Instruct Models

- **OpenAI**
  - `text-davinci-003`
  - `gpt-3.5-turbo`
  - `gpt-4`
- **Anthropic**
  - Claude 1, Claude 2, Claude 3
- **Google DeepMind**
  - Gemini Pro
  - Gemini Ultra

- **Meta**
    - LLaMA 2 Chat
    - LLaMA 3 Chat
- **Mistral**
    - Mistral Instruct v0.2
    - Mixtral (Mixture of Experts) Instruct
- **Cohere**
    - Command R
    - Command R+
    - Command Light
- **Amazon**
    - Titan Text Lite
    - Titan Text Express
- **TII (Technology Innovation Institute)**
    - Falcon-7B-Instruct
    - Falcon-40B-Instruct
- **MosaicML**
    - MPT-7B-Instruct
- **Databricks**
    - Dolly v2
- **Open Source Community**
    - Alpaca
    - Vicuna
    - OpenChat
    - Nous-Hermes

- Zephyr
- Orca
- Baize

These instruct models are foundational for building intelligent agents and chat systems that can understand tasks, interact with tools, and produce reliable, context-aware responses. Selecting the right instruct model depends on the use case, performance requirements, and whether the deployment is cloud-based or local.

Summary

LangChain's advanced components—like memory modules, embedding models, document loaders, retrievers, and agents—offer powerful building blocks for creating intelligent, context-aware applications. These tools allow developers to go beyond simple prompt chaining, enabling capabilities such as conversational memory, dynamic reasoning, document search, and seamless integration with external tools and data sources. With these components, LangChain empowers developers to build robust, adaptive systems tailored to real-world needs.

In the next chapter, we'll explore how to apply these features by developing a variety of advanced, practical applications. From personal assistants and customer support bots to document Q&A systems and multiagent workflows, we'll walk through real use cases that demonstrate LangChain's full potential in action.

CHAPTER 3

Building Advanced Applications Powered by LLMs with LangChain and Python

In the rapidly evolving world of artificial intelligence, large language models (LLMs) have emerged as powerful engines that drive innovative applications, transforming the way we interact with technology. This chapter is a deep dive into the sophisticated strategies and techniques that harness the full potential of these models. Here, we explore how to go beyond basic implementations and craft complex, robust systems that leverage LLMs to address real-world challenges.

At the core of our discussion is LangChain—a versatile framework designed to streamline the integration of LLMs into advanced application architectures. LangChain provides a modular and extensible environment that simplifies the orchestration of language model tasks, allowing developers to build applications that can manage multistep reasoning, handle dynamic interactions, and maintain contextual continuity across extended dialogues. Coupled with the power and flexibility of Python, LangChain equips you with the tools to push the boundaries of what is possible in modern software development.

In this chapter, we begin by revisiting the foundational concepts behind LLMs, setting the stage for a more nuanced understanding of their capabilities and limitations. We then transition into an exploration of LangChain's architecture, examining its key components and how they work together to facilitate advanced workflows. Through

D. Grigorov, *Intermediate Python and Large Language Models*, https://doi.org/10.1007/979-8-8688-1475-4_3

detailed examples and hands-on exercises, you will learn how to implement complex pipelines that integrate external data sources, manage iterative processing, and optimize performance under demanding conditions.

As we progress, the focus shifts to the practical challenges encountered when building advanced LLM-powered applications. We will discuss strategies for fine-tuning model behavior, mitigating errors, and ensuring scalability in production environments. Special emphasis is placed on designing systems that not only perform efficiently but also maintain high levels of reliability and security. This chapter provides insights into best practices for monitoring, debugging, and continuously improving your applications, ensuring that they remain robust in the face of evolving requirements and emerging threats.

Moreover, we will highlight cutting-edge use cases that demonstrate the transformative impact of advanced LLM applications. From intelligent virtual assistants that seamlessly manage complex conversations to automated content generation systems capable of nuanced analysis and synthesis, you will see how the principles discussed can be applied to a diverse array of challenges. By dissecting these real-world examples, you will gain a deeper appreciation of the potential for innovation when combining LangChain's orchestration capabilities with Python's rich ecosystem.

Whether you are an experienced developer looking to elevate your skill set or a curious practitioner eager to explore the next frontier in AI application development, this chapter is designed to equip you with the knowledge and tools necessary to build advanced, production-grade systems. By the end of our journey, you will have a comprehensive understanding of how to harness LLMs effectively, enabling you to create applications that are not only intelligent but also adaptive, scalable, and ready for the challenges of tomorrow.

In the next pages, we will

- **Build a YouTube Video Summarizer**
    - Automatically transcribe and summarize long YouTube videos for quick content digestion
- **Create a GitHub repository chatbot**
    - Interact with code bases conversationally by indexing and querying repository files

- **Develop a financial report analysis tool**
    - Analyze and extract insights from financial documents using AI-driven retrieval and Q&A
- **Enhance blog content with Google Search**
    - Use LLMs and live web data to intelligently expand and enrich blog posts
- **Automate YouTube scriptwriting**
    - Generate structured, engaging scripts from video transcripts with GPT models
- **Design an AI-powered email generator**
    - Instantly craft professional, personalized email responses using a customizable prompt
- **Analyze CSV data with AI assistance**
    - Load, summarize, and visualize datasets with natural language commands and visual tools

Each app is presented with

- Step-by-step implementation instructions
- Clear explanations of LangChain and OpenAI integrations
- Tips for optimizing performance, usability, and scalability

By the end of this chapter, you'll be equipped to build your own intelligent, production-ready applications with Python and LLMs.

App 1: YouTube Video Summarizer

In the digital age, YouTube has become a vast repository of knowledge, offering millions of videos on various topics, from educational lectures to industry insights. However, watching long videos to extract key information can be time-consuming. This is where a **YouTube Video Summarizer with LangChain** comes into play.

A **YouTube Video Summarizer** is an AI-powered tool that automatically transcribes and summarizes YouTube videos, providing users with a concise and structured overview of the content. By leveraging **LangChain**, a framework designed for building applications with **large language models (LLMs)**, the summarizer efficiently processes video transcripts and distills essential insights.

This tool utilizes **natural language processing (NLP)** to extract meaningful information, making it easier for users to grasp the key points of a video in seconds rather than minutes or hours. Whether you are a researcher, student, or content creator, a **YouTube Video Summarizer with LangChain** enhances productivity by offering quick and accurate video summaries, helping you stay informed without watching entire videos.

How to Build the App

Step 1: Get Your OpenAI API Key

You need to get your OpenAI API key here: `https://platform.openai.com/settings/organization/api-keys`.

Step 2: Run the Following Commands

Run the following commands in your environment—in our case, this is Google Colab—to install libraries needed:

```
!pip install langchain==0.3.23 activeloop-deeplake==3.9.5 openai==1.3.12
tiktoken==0.7.0 langchain-openai==0.3.12
```

```
!pip install -q yt_dlp
!pip install -q git+https://github.com/openai/whisper.git
Also install ffmpeg:
conda install ffmpeg
```

```
For Google Colab:
```

```
!apt-get update -qq && apt-get install -y ffmpeg
```

Note Due to the security precautions taken by YouTube, you need to download a browser extension to download the cookies from your browser for your desired video.

For example, for Chrome you can use: `https://chromewebstore.google.com/detail/get-cookiestxt-locally/cclelndahbckbenkjhflpdbgdldlbecc`. Then, upload the cookie file to your Google Colab files.

Step 3: Execute the Following Command

Execute the following command to download your video with your desired file name:

```
!yt-dlp --cookies cookies.txt -f "bestvideo[ext=mp4]+bestaudio[ex
t=m4a]/best[ext=mp4]" -o "my_video.mp4" "https://www.youtube.com/
watch?v=Gx5qb1uHss4"
```

Explanation of the command:

- `-f "bestvideo[ext=mp4]+bestaudio[ext=m4a]/best[ext=mp4]"`
    - This ensures that yt-dlp downloads the **best quality MP4 video** and **best M4A audio** and then merges them.
    - If a single **MP4 format video with audio** is available, it downloads that.
- `-o "my_video.mp4"`
    - This sets the output filename as `my_video.mp4`.
- `--cookies cookies.txt`
    - This allows yt-dlp to use authentication for downloading restricted videos.

Step 4: Import the Whisper Model and Process the Video

```
import whisper

model = whisper.load_model("base")
result = model.transcribe("my_video.mp4")
print(result['text'])
```

This Python script utilizes OpenAI's Whisper model for automatic speech recognition (ASR) to transcribe the audio from a given video file. It begins by importing the Whisper library, which is responsible for handling the transcription process. Next, it loads a pretrained Whisper model, specifically the "base" version, which is a lightweight model compared to larger variants like "medium" or "large." If the model is not already available locally, it will be automatically downloaded from OpenAI's servers.

Once the model is loaded, the script processes the specified video file by extracting its audio and converting the spoken content into text. Finally, the transcribed text is extracted from the result and printed to the console.

Step 5: Read the Written Content in a File

```
with open ('text.txt', 'w') as file:
   file.write(result['text'])
```

Step 6: Use LangChain to Split a Text File into Smaller Chunks

Let's use LangChain to split a text file into smaller chunks for further processing, such as feeding into a language model.

```
from langchain.text_splitter import RecursiveCharacterTextSplitter
from langchain.docstore.document import Document

text_splitter = RecursiveCharacterTextSplitter(
    chunk_size=1000, chunk_overlap=0, separators=[" ", ",", "\n"]

)
```

```
with open('text.txt') as f:
    text = f.read()

texts = text_splitter.split_text(text)
docs = [Document(page_content=t) for t in texts[:4]]
```

Step-by-Step Explanation

1. **Import necessary modules from LangChain**
    - **RecursiveCharacterTextSplitter**: A utility for breaking long texts into smaller pieces while trying to maintain meaning
    - **Document**: A simple wrapper around text content, useful when dealing with large documents
2. **Initialize the text splitter**
    - The RecursiveCharacterTextSplitter is configured with
        - chunk_size=1000: Each chunk will have a maximum of **1000 characters.**
        - chunk_overlap=0: No overlap between chunks.
        - separators=[“ ”, “,”, “\n”]: The text will be split **preferentially** at spaces, then commas, then newlines.
3. **Read text from a file (text.txt)**
    - The text content is loaded into memory as a single string.
4. **Split the text into smaller chunks**
    - split_text(text): The loaded text is split into multiple **chunks** of up to 1000 characters each.
    - The splitting occurs **recursively**, prioritizing the given separators.
5. **Convert the first four chunks into Document objects**
    - The first four chunks (texts[:4]) are wrapped in Document objects.
    - Each Document stores its corresponding chunk as page_content.

Step 7: Summarize the Preprocessed Content

```
from langchain.chains.summarize import load_summarize_chain
import textwrap
from langchain_openai import OpenAI
from langchain.chains import LLMChain
from langchain.chains.summarize import load_summarize_chain
from langchain.prompts import PromptTemplate

# Initialize OpenAI LLM
llm = OpenAI(api_key="sk-proj-MiOxm6IEXSZ8ULyof8caT3BlbkFJNHaISzb3nz3hsau8t
qyn", model="gpt-3.5-turbo-instruct", temperature=0)
chain = load_summarize_chain(llm, chain_type="map_reduce")

output_summary = chain.invoke(docs)
wrapped_text = textwrap.fill(output_summary["output_text"], width=100)
```

Step-by-step execution:

1. **Import necessary libraries**
    - `load_summarize_chain`: A utility to create a summarization pipeline
    - `textwrap`: Used to format the output text
    - `OpenAI`: Initializes OpenAI's GPT model for text processing
    - `LLMChain`: A generic LangChain wrapper for using LLMs
    - `PromptTemplate`: Allows customization of prompts for the LLM
2. **Initialize OpenAI language model**
    - The `OpenAI` LLM is initialized with
        - The **GPT-3.5 Turbo Instruct** model
        - **Temperature** = **0**, ensuring deterministic (consistent) responses
        - The **API key** (which should be kept secret)

3. **Load a summarization chain**

   - The `load_summarize_chain(llm, chain_type="map_reduce")` initializes a **two-step summarization pipeline**:

     - **Map Stage**: Each document is summarized individually.

     - **Reduce Stage**: The individual summaries are combined into a final, coherent summary.

4. **Invoke the summarization chain**

   - `chain.invoke(docs)`: The chain takes `docs` (a list of `Document` objects) and processes them.

   - `output_summary["output_text"]`: Extracts the final summarized text.

5. **Format the summary output**

   - `textwrap.fill(output_summary["output_text"], width=100)` wraps the summary text so that lines do not exceed **100 characters** in width.

Step 8: Define a Prompt Template Using LangChain's PromptTemplate

The following code defines a prompt template using LangChain's PromptTemplate to structure input for an LLM (large language model), such as OpenAI's GPT models:

```
prompt_template = """Write a concise bullet point summary of the following:

{text}

CONSCISE SUMMARY IN BULLET POINTS:"""

BULLET_POINT_PROMPT = PromptTemplate(template=prompt_template,
                                    input_variables=["text"])
```

Step 9: Summarization Pipeline

```
chain = load_summarize_chain(llm,
                             chain_type="stuff",
                             prompt=BULLET_POINT_PROMPT)

output_summary = chain.run(docs)

wrapped_text = textwrap.fill(output_summary,
                             width=1000,
                             break_long_words=False,
                             replace_whitespace=False)
print(wrapped_text)
```

This code sets up a summarization pipeline using LangChain and an LLM, such as OpenAI's GPT-3.5 Turbo. The process begins by initializing a summarization chain with a specific configuration. The chain type is set to process all text at once, rather than breaking it into smaller sections. A custom prompt template is used to instruct the model to generate a structured bullet-point summary.

Once the summarization chain is created, it is executed using a list of documents as input. The chain processes the text and generates a concise summary. The resulting summary is then formatted for better readability by ensuring that lines do not exceed a certain width, words are not split across lines, and whitespace formatting is preserved. Finally, the formatted summary is displayed as output.

App 2: Chat with a GitHub Repository

This Python application enables users to interact with a GitHub repository using natural language. It utilizes **LangChain**, **OpenAI embeddings**, and **FAISS vector storage** to process and retrieve relevant code snippets, documentation, and README contents from a repository.

How It Works

1. **Fetches Repository Data**: Uses the GitHub API to retrieve all files in the repository
2. **Embeds and Indexes Content**: Converts text into embeddings for efficient search
3. **Conversational Retrieval**: Allows users to ask questions and get relevant information
4. **Memory Support**: Maintains context in ongoing conversations for a better chat experience

Step 1: Select a GitHub Repository and Download It As Zip

For example, `https://github.com/milaan9/07_Python_Advanced_Topics`, and get its username and repo name—in this case, its **milaan9**, and the name of the repo is **07_Python_Advanced_Topics**. Or in other words, you can find it in the form `https://github.com/{user_name}/{repo_name}`.

Step 2: Install All Libraries Required

```
!pip install langchain==0.3.23 openai==1.3.12 faiss-cpu==1.8.0
tiktoken==0.7.0 requests==2.31.0 python-dotenv==1.0.1

!pip install langchain-community==0.3.23
```

Step 3: Import the Libraries and Obtain the Needed API Keys

For OpenAI: `https://platform.openai.com/api-keys`

For GitHub: `https://github.com/settings/tokens`

```
import os
import requests
from dotenv import load_dotenv
from langchain.embeddings import OpenAIEmbeddings
```

```
from langchain.vectorstores import FAISS
from langchain.text_splitter import RecursiveCharacterTextSplitter
from langchain.llms import OpenAI
from langchain.chains import ConversationalRetrievalChain
from langchain.memory import ConversationBufferMemory
from langchain.chat_models import ChatOpenAI

load_dotenv()

GITHUB_TOKEN = os.getenv("GITHUB_TOKEN")
OPENAI_API_KEY = os.getenv("OPENAI_API_KEY")

headers = {"Authorization": f"token {GITHUB_TOKEN}"}
```

Step 4: Get Repository Content

The function is designed to **retrieve the contents** of a GitHub repository using the GitHub API. It allows you to access files and directories within a repository.

1. **It accepts three inputs**
    - The **owner** of the repository (a username or organization name).
    - The **repository name** to fetch data from.
    - An optional **path** specifying a file or folder within the repository. If no path is provided, it retrieves the root directory.
2. **It builds a URL**
    - The function constructs a web address following GitHub's API format. This URL points to the requested repository and its contents.
3. **It sends a request to GitHub**
    - A request is made to GitHub's servers to fetch the contents of the specified file or folder.
4. **It checks for a successful response**
    - If GitHub responds successfully, the function extracts and returns the content in a structured format (as data).

- If the request fails (e.g., due to incorrect repository details, permission issues, or rate limits), an error message is displayed, and an empty response is returned.

5. **It handles different repository structures**

   - If the request targets a directory, the function retrieves a list of its files and subdirectories.
   - If it targets a file, it fetches the file's content and relevant metadata.

```
def get_repo_contents(owner, repo, path=""):
    url = f"https://api.github.com/repos/{owner}/{repo}/contents/{path}"
    response = requests.get(url, headers=headers)
    if response.status_code == 200:
        return response.json()
    else:
        print("Error fetching repo contents:", response.json())
        return []
```

Step 5: Fetch All Files

```
def fetch_all_files(owner, repo, path="", collected_files=None):
    if collected_files is None:
        collected_files = {}

    contents = get_repo_contents(owner, repo, path)

    for item in contents:
        if item["type"] == "file":
            file_content = requests.get(item["download_url"],
            headers=headers).text
            collected_files[item["path"]] = file_content
        elif item["type"] == "dir":
            fetch_all_files(owner, repo, item["path"], collected_files)

    return collected_files
```

In this function, we **retrieve all files** from a GitHub repository, including those inside subdirectories. It works by **recursively** navigating through the repository's structure and collecting file contents.

Step-by-Step Explanation

1. **It initializes a dictionary to store files**
    - If no dictionary is provided, an empty one is created to store file paths and their contents.
2. **It retrieves repository contents**
    - The function calls another function to fetch the list of files and folders at a given location in the repository.
    - If no specific path is provided, it starts from the root directory.
3. **It loops through each item in the retrieved list**
    - If the item is a **file**, the function
        - Downloads its content from GitHub
        - Stores the file's path as a key and its content as a value in the dictionary
    - If the item is a **directory**, the function
        - Calls itself again (recursion), using the directory's path as the new starting point
        - This ensures all nested files and folders are processed
4. **It returns a dictionary containing all files**
    - After processing all files and directories, the function returns a dictionary where
        - Each **key** represents a file's path within the repository
        - Each **value** contains the corresponding file's content

Step 6: Creating a Searchable Database

```
def create_vector_db(files):
    texts = []
    for path, content in files.items():
        texts.append(f"### {path}\n{content}")

    text_splitter = RecursiveCharacterTextSplitter(chunk_size=1000, chunk_
    overlap=100)
    docs = text_splitter.create_documents(texts)

    embeddings = OpenAIEmbeddings()
    vectorstore = FAISS.from_documents(docs, embeddings)
    return vectorstore
```

In this function, we **create a searchable database** from a collection of files by converting their contents into numerical representations (embeddings) that allow efficient retrieval.

Step-by-Step Explanation

1. **It prepares the text data**
    - The function starts with an empty list to store text data.
    - It loops through each file in the input dictionary, which contains file paths as keys and their contents as values.
    - Each file's content is formatted with a header that includes the file path, ensuring that file names remain associated with their contents.
2. **It splits the text into chunks**
    - Since some files may be large, the function breaks them into smaller chunks.
    - A text-splitting tool is used to divide the text while maintaining some overlap between chunks to preserve context.

- This ensures that each chunk is not too large for processing while still making sense when analyzed.

3. **It converts text into embeddings**

   - The function uses an embedding model to transform the text chunks into **numerical vectors**.
   - These embeddings capture the **semantic meaning** of the text, making it possible to search for similar content based on meaning rather than exact words.

4. **It stores the embeddings in a searchable database**

   - A specialized database (FAISS) is used to store these embeddings efficiently.
   - FAISS allows quick searching and retrieval of relevant text based on similarity to a given query.

5. **It returns the searchable database**

   - The final result is a structured vector database that enables quick searches for relevant file contents.

Step 7: Creating the Actual Chatting Feature Function

```
def chat_with_repo(owner, repo):
    files = fetch_all_files(owner, repo)
    vectorstore = create_vector_db(files)
    retriever = vectorstore.as_retriever()

    memory = ConversationBufferMemory(memory_key="chat_history", return_
    messages=True)
    chat = ConversationalRetrievalChain.from_llm(
        llm=ChatOpenAI(model_name="gpt-4", temperature=0.5),
        retriever=retriever,
        memory=memory
    )

    print("Chat with the GitHub repository! Type 'exit' to quit.")
```

```
while True:
    query = input("You: ")
    if query.lower() == "exit":
        break
    response = chat.run(query)
    print("Bot:", response)
```

It allows a user to **interactively chat** with a GitHub repository by retrieving relevant information from its contents.

Step-by-Step Explanation

1. **It fetches all files from the repository**
    - The function retrieves the entire repository's contents, including files in subdirectories.
    - This ensures that all text-based content is available for processing.
2. **It creates a searchable vector database**
    - The fetched files are processed and converted into a **vector database**.
    - This allows the chatbot to **search for relevant information** efficiently.
3. **It sets up a retriever**
    - A retriever is initialized from the vector database.
    - The retriever helps the chatbot find relevant file contents when answering questions.
4. **It initializes memory for conversations**
    - A memory module is added to keep track of past interactions.
    - This allows the chatbot to maintain context throughout the conversation.

5. **It creates a conversational AI model**
    - A **GPT-4 language model** is loaded to generate responses.
    - The model uses the retriever to find relevant repository content when answering questions.
6. **It starts an interactive chat**
    - The function displays a message prompting the user to start chatting.
    - It continuously takes user input, processes it, and provides responses.
    - If the user types `"exit"`, the chat ends.

Output:

```
Enter GitHub owner/org: milaan9
Enter repository name: 07_Python_Advanced_Topics
Chat with the GitHub repository! Type 'exit' to quit.
You: What are the advanced topics in the repo?
Bot: The advanced topics in the repository are:

1. Python Iterators
2. Python Generators
3. Python Closure
4. Python Decorators
   - Python args and kwargs
5. Python Property
6. Python RegEx
```

App 3: Financial Report Analysis App

This app is designed to streamline financial data analysis by leveraging **AI-powered document retrieval** and **natural language processing**. Built with **LangChain, FAISS, and OpenAI models**, it allows users to efficiently search and analyze financial reports, specifically those from Amazon, but it can be adapted for any financial documents.

Key Features

- **Automated PDF Parsing**: Extracts financial data from multiple PDF reports.
- **AI-Driven Search**: Uses advanced embeddings and retrieval mechanisms to provide quick answers to financial queries.
- **Efficient Data Management**: Utilizes FAISS for fast and scalable vector-based document retrieval.
- **Conversational Querying**: Enables users to ask specific questions (e.g., *"What was Amazon's revenue in Q3 2021?") and* get direct answers.

This tool is ideal for **financial analysts, researchers, and business professionals** who need instant insights from large datasets without manually scanning through reports. Whether you're tracking revenue trends, identifying financial performance, or analyzing key business metrics, this app provides a **seamless and intelligent solution** to financial document analysis.

Step 1: Install All Required Libraries

```
!pip3 install langchain faiss-cpu pypdf openai tiktoken langchain-openai
langchain-community
```

Step 2: Set Up OpenAI API Key and Add It to the Code

```
# Set API keys (Use environment variables for security)
import os
os.environ["OPENAI_API_KEY"] = "Your OpenAI Key"
```

Step 3: Import All Required Libraries

```
from langchain_openai import OpenAIEmbeddings
from langchain.vectorstores import FAISS
from langchain.text_splitter import CharacterTextSplitter
from langchain_openai import OpenAI
```

```
from langchain.chains import RetrievalQA
from langchain_openai import ChatOpenAI
from langchain.document_loaders import PyPDFLoader

import requests
import tqdm
from typing import List
```

Step 4: Process Financial Reports

```
import requests
import tqdm
from typing import List

# financial reports of amamzon, but can be replaced by any URLs of pdfs
urls = ['https://s2.q4cdn.com/299287126/files/doc_financials/Q1_2018_-_8-K_
Press_Release_FILED.pdf',
        'https://s2.q4cdn.com/299287126/files/doc_financials/Q2_2018_
        Earnings_Release.pdf',
        'https://s2.q4cdn.com/299287126/files/doc_news/archive/Q318-Amazon-
        Earnings-Press-Release.pdf',
        'https://s2.q4cdn.com/299287126/files/doc_news/archive/AMAZON.COM-
        ANNOUNCES-FOURTH-QUARTER-SALES-UP-20-TO-$72.4-BILLION.pdf',
        'https://s2.q4cdn.com/299287126/files/doc_financials/Q119_Amazon_
        Earnings_Press_Release_FINAL.pdf',
        'https://s2.q4cdn.com/299287126/files/doc_news/archive/Amazon-
        Q2-2019-Earnings-Release.pdf',
        'https://s2.q4cdn.com/299287126/files/doc_news/archive/Q3-2019-
        Amazon-Financial-Results.pdf',
        'https://s2.q4cdn.com/299287126/files/doc_news/archive/Amazon-
        Q4-2019-Earnings-Release.pdf',
        'https://s2.q4cdn.com/299287126/files/doc_financials/2020/Q1/AMZN-
        Q1-2020-Earnings-Release.pdf',
        'https://s2.q4cdn.com/299287126/files/doc_financials/2020/q2/
        Q2-2020-Amazon-Earnings-Release.pdf',
```

```
        'https://s2.q4cdn.com/299287126/files/doc_financials/2020/q4/
        Amazon-Q4-2020-Earnings-Release.pdf',
        'https://s2.q4cdn.com/299287126/files/doc_financials/2021/q1/
        Amazon-Q1-2021-Earnings-Release.pdf',
        'https://s2.q4cdn.com/299287126/files/doc_financials/2021/q2/AMZN-
        Q2-2021-Earnings-Release.pdf',
        'https://s2.q4cdn.com/299287126/files/doc_financials/2021/q3/
        Q3-2021-Earnings-Release.pdf',
        'https://s2.q4cdn.com/299287126/files/doc_financials/2021/q4/
        business_and_financial_update.pdf',
        'https://s2.q4cdn.com/299287126/files/doc_financials/2022/q1/
        Q1-2022-Amazon-Earnings-Release.pdf',
        'https://s2.q4cdn.com/299287126/files/doc_financials/2022/q2/
        Q2-2022-Amazon-Earnings-Release.pdf',
        'https://s2.q4cdn.com/299287126/files/doc_financials/2022/q3/
        Q3-2022-Amazon-Earnings-Release.pdf',
        'https://s2.q4cdn.com/299287126/files/doc_financials/2022/q4/
        Q4-2022-Amazon-Earnings-Release.pdf'
        ]

def load_reports(urls: List[str]) -> List[str]:
    """ Load pages from a list of urls"""
    pages = []

    for url in tqdm.tqdm(urls):
        r = requests.get(url)
        path = url.split('/')[-1]
        with open(path, 'wb') as f:
            f.write(r.content)
        loader = PyPDFLoader(path)
        local_pages = loader.load_and_split()
        pages.extend(local_pages)
    return pages

pages = load_reports(urls)
```

The code downloads Amazon's financial reports in PDF format, extracts their text, and stores the content in a list. It starts by iterating through a predefined list of URLs, downloading each PDF using the `requests` library, and saving the files locally. Then, it processes each file with `PyPDFLoader` to extract and split the text into pages, which are appended to a list. The `tqdm` library provides a progress bar to track the downloading process. Finally, the extracted text from all PDFs is stored in the `pages` list for further analysis. However, the script is missing an import for `PyPDFLoader`, which would cause an error unless added manually. Additionally, the saved PDFs are not deleted after extraction.

Step 5: Preparing and Indexing Text Data for Efficient Retrieval Using AI-Powered Search and Question Answering (QA)

1. Splitting the Extracted Text into Smaller Chunks

```
text_splitter = CharacterTextSplitter(chunk_size=1000, chunk_overlap=0)
texts = text_splitter.split_documents(pages)
```

- **What it does:**
    - The extracted text from financial reports (stored in `pages`) is often long and unstructured.
    - The `CharacterTextSplitter` takes these texts and **breaks them into smaller chunks** of 1000 characters each (`chunk_size=1000`).
    - There is **no overlap** between chunks (`chunk_overlap=0`), meaning each piece of text is distinct.
- **Why it's needed:**
    - Splitting text into smaller sections allows for **better indexing** and **faster retrieval** when querying later.

2. Generating Text Embeddings

```
embeddings = OpenAIEmbeddings()
```

- **What it does:**
  - Uses **OpenAI's embedding model** to convert text chunks into **numerical vectors.**
  - These vectors **capture the meaning of the text** so they can be efficiently searched.
- **Why it's needed:**
  - A numerical representation (embedding) allows us to perform **semantic search**—meaning we can find relevant information **even if the search query does not exactly match the words in the document.**

3. Storing and Indexing the Text Chunks in a FAISS Database

```
db = FAISS.from_documents(texts, embeddings)
```

- **What it does:**
  - Uses **FAISS (Facebook AI Similarity Search)**, a powerful **vector database**, to store the generated embeddings.
  - FAISS allows for **efficient and fast searching** of similar text embeddings.
- **Why it's needed:**
  - Instead of searching through raw text, we **search through embeddings**, making retrieval **faster and more accurate**.

4. Setting Up the AI-Powered Retrieval and QA System

```
qa = RetrievalQA.from_chain_type(llm=ChatOpenAI(model='gpt-3.5-turbo'),
chain_type='stuff', retriever=db.as_retriever())
```

- **What it does:**
    - Uses **ChatGPT (gpt-3.5-turbo)** to answer user queries based on the indexed documents.
    - The `retriever=db.as_retriever()` ensures that **relevant text chunks are retrieved** from FAISS before being processed by the AI.
    - The `chain_type='stuff'` method combines the retrieved text into a single response.
- **Why it's needed:**
    - This setup allows the app to **answer questions** like *"What was Amazon's revenue in Q3 2021?"* based on financial reports **without needing a human to manually search the documents**.

Step 6: Ask a Question

```
qa.invoke("What is the revenue in 2021 Q3?")
```

Output:

```
{'query': 'What is the revenue in 2021 Q3?',
 'result': 'The revenue for Amazon in 2021 Q3 was $110.8 billion.'}
```

App 4: Automate and Enhance Your Blog Posts with LangChain and Google Search

Artificial intelligence is transforming the field of copywriting by acting as a powerful writing assistant. Modern language models can detect grammar and spelling errors, adjust tone, summarize content, and even expand text. However, these models sometimes lack the deep domain expertise needed to provide high-quality extensions for specific topics.

In this lesson, we'll guide you through building an application that seamlessly enhances text sections. The process starts by prompting a language model (such as ChatGPT) to generate relevant search queries based on the existing content. These queries are then used with the Google Search API to retrieve authoritative information from the Web. Finally, the most relevant results are provided as context to the model, allowing it to generate more accurate and well-informed content suggestions.

Step 1: Install All Required Libraries

```
!pip install langchain==0.0.208 deeplake==3.9.27 openai==0.27.8 tiktoken
!pip install -q newspaper3k==0.2.8 python-dotenv
!pip install lxml_html_clean
```

Step 2: Define Three Variables—Title, Text All, and Text to Change

Here, we have three variables that store an article's title and content (`text_all`) from *Artificial Intelligence News*. Additionally, the `text_to_change` variable identifies the specific section of the text that we want to expand. These constants serve as reference points and will remain unchanged throughout the lesson.

```
title = "OpenAI Chief: AI Oversight 'Crucial' for Future Innovation"

text_all = """ Altman underscored the immense potential of AI advancements
such as ChatGPT and DALL-E 2 in tackling global challenges like climate
change and disease research. However, he also cautioned against the
unchecked proliferation of increasingly capable AI models. To address
these concerns, he suggested that governments explore regulatory measures,
including licensing frameworks and stringent testing protocols for high-
capability AI systems. Altman reaffirmed OpenAI's dedication to responsible
AI development, ensuring rigorous evaluations before deploying new
technologies. Senators Josh Hawley and Richard Blumenthal acknowledged AI's
disruptive impact and the necessity of understanding its ramifications on
elections, employment, and national security. To illustrate AI's power,
Blumenthal played an audio clip generated by an AI voice cloning system
trained on his past speeches. He raised alarms over AI-related threats such
as misinformation, manipulated media, discrimination, cyber harassment, and
identity fraud. Additionally, he warned about the displacement of human
workers amid an AI-driven economic transformation."""

text_to_change = """ Senators Josh Hawley and Richard Blumenthal
```

```
acknowledged AI's disruptive impact and the necessity of understanding
its ramifications on elections, employment, and national security. To
illustrate AI's power, Blumenthal played an audio clip generated by an AI
voice cloning system trained on his past speeches."""
```

We start by generating potential search queries from the paragraph we want to expand. These queries are then used to retrieve relevant documents from a search engine (such as Bing or Google Search), which are subsequently broken down into smaller chunks. Next, we compute embeddings for these chunks and store both the chunks and their embeddings in a Deep Lake dataset. Finally, the most relevant chunks are retrieved from Deep Lake based on their similarity to the original paragraph. These retrieved chunks are then incorporated into a prompt to enhance the paragraph with additional context and information.

Step 3: Define Your API Keys

```
# Set API keys (Use environment variables for security)
import os
os.environ["OPENAI_API_KEY"] = "Your API Key"
os.environ["GOOGLE_API_KEY"] = "Your API Key"
os.environ["GOOGLE_CSE_ID"] = "Your ID"
```

How to Get Your API Keys and ID

To use the Google Search API, we first need to set up an API key and a custom search engine. Start by navigating to the Google Cloud Console and creating a project, and then, enable the **Custom Search API** under **Enable APIs and Services** (Google will provide instructions if necessary). After that, generate an API key by clicking **CREATE CREDENTIALS** at the top and selecting **API KEY**.

Once these steps are complete, configure the environment variables "**GOOGLE_CSE_ID**" and "**GOOGLE_API_KEY**", allowing the Langchain Google wrapper to connect seamlessly with the API.

Next, go to the Programmable Search Engine dashboard: `https://programmablesearchengine.google.com/controlpanel/create`, create a custom search engine, and ensure that the "Search the entire web" option is selected. The search engine ID will be displayed in the Details section.

Go to the page with all search engines: `https://programmablesearchengine.google.com/controlpanel/all`, and click the one you have just created. Then, copy the Search engine ID.

Step 4: Generate Search Results

The following code leverages OpenAI's ChatGPT model to analyze an article and generate three relevant search queries. It begins by defining a prompt that instructs the model to suggest Google search queries for gathering more information on the topic. The "LLMChain" component connects the "ChatOpenAI" model with the "ChatPromptTemplate", forming a structured pipeline for interacting with the model.

Once the response is received, the code splits it by newline and removes the initial characters to extract the search queries. This approach works because the API was instructed to format each query as a new line starting with "-". (Alternatively, the same result can be achieved using the "OutputParser" class.)

Before executing the code, ensure that your OpenAI API key is stored in the "OPENAI_API_KEY" environment variable.

```
from langchain.chat_models import ChatOpenAI
from langchain.chains import LLMChain
from langchain.prompts import PromptTemplate
from langchain.prompts.chat import (
    ChatPromptTemplate,
    HumanMessagePromptTemplate,
)

template = """ You are an exceptional copywriter and content creator.

You're reading an article with the following title:
----------------
{title}
----------------

You've just read the following piece of text from that article.
----------------
{text_all}
```

```
----------------

Inside that text, there's the following TEXT TO CONSIDER that you want to
enrich with new details.
----------------
{text_to_change}
----------------

What are some simple and high-level Google queries that you'd do to search
for more info to add to that paragraph?
Write 3 queries as a bullet point list, prepending each line with -.
"""

human_message_prompt = HumanMessagePromptTemplate(
    prompt=PromptTemplate(
        template=template,
        input_variables=["text_to_change", "text_all", "title"],
    )
)
chat_prompt_template = ChatPromptTemplate.from_messages([human_message_prompt])

# Before executing the following code, make sure to have
# your OpenAI key saved in the “OPENAI_API_KEY” environment variable.
chat = ChatOpenAI(model_name="gpt-4o-mini", temperature=0.5)
chain = LLMChain(llm=chat, prompt=chat_prompt_template)

response = chain.run({
    "text_to_change": text_to_change,
    "text_all": text_all,
    "title": title
})

response_queries = [line[2:] for line in response.split("\n")]
queries = [item.replace('"', "") for item in response_queries]
print(queries)
```

Output: ['impact of AI on elections and democracy 2023 ', 'AI voice cloning technology examples and implications ', 'AI effects on employment and workforce displacement 2023 ']

Step 5: Get Search Results

To use the Google Search API, we first need to set up an API key and a custom search engine. Start by navigating to the **Google Cloud Console**, and then, generate an API key by clicking **CREATE CREDENTIALS** at the top and selecting **API KEY**. Next, go to the **Programmable Search Engine** dashboard, and ensure that the **"Search the entire web"** option is enabled. The search engine ID will be displayed in the Details section.

Additionally, you may need to enable the **"Custom Search API"** under **Enable APIs and Services** (Google will provide further instructions if required). Once these steps are complete, configure the environment variables GOOGLE_CSE_ID and GOOGLE_API_KEY, enabling the Google wrapper to interact with the API.

The next step is to use the generated search queries from the previous section to retrieve relevant sources from Google. The **LangChain** library offers the GoogleSearchAPIWrapper, which handles search queries and retrieves results. To process the results efficiently, we define a function using the top_n_results parameter.

Then, the Tool class creates a wrapper around this function, making it compatible with AI agents so they can interact with external data sources. We request only the **top five search results** and then concatenate them for each query into the all_results variable for further processing.

```
from langchain.tools import Tool
from langchain.utilities import GoogleSearchAPIWrapper

# Initialize the Google Search API Wrapper
search = GoogleSearchAPIWrapper()
TOP_N_RESULTS = 5

def top_n_results(query):
    """Fetch top N search results for a given query."""
    results = search.results(query, TOP_N_RESULTS)
    if not results:
        return [{"Result": "No good Google Search Result was found"}]
    return results

# Define the search tool
search_tool = Tool(
```

```
    name="Google Search",
    description="Search Google for recent results.",
    func=top_n_results
)

# Sample queries list
queries = ['Senators Josh Hawley Richard Blumenthal AI regulation
statements', 'impact of AI on elections jobs security 2023', 'AI voice
cloning technology examples implications']
all_results = []

# Run search for each query
for query in queries:
    try:
        results = search_tool.run(query)
        all_results.extend(results)
    except Exception as e:
        all_results.append({"Error": str(e)})

# Print all collected search results
print(all_results)
```

The "all_results" variable may contain a different number of web addresses—derived from three search queries generated by ChatGPT, each returning the top five Google search results. However, using all retrieved content as context in our application is not an optimal approach due to technical, financial, and contextual constraints.

First, **language models (LLMs) have input length limitations**, typically ranging from **2K to 4K tokens**, depending on the model. While alternative chain types can help bypass this constraint, staying within the model's token window is often more efficient and produces better results.

Second, **cost considerations** come into play. The more text we send to the API, the higher the cost. Although splitting prompts into multiple chains is an option, we must be mindful that API pricing is based on token usage, making excessive input size financially inefficient.

Finally, **contextual relevance matters**. The retrieved search results will likely contain overlapping or similar information. Instead of using all results indiscriminately, selecting the most relevant ones ensures a more focused and meaningful expansion of the content.

Output: [{'title': '[2023-09-08] Blumenthal & Hawley Announce Bipartisan Framework ...', 'link': 'https://www.blumenthal.senate.gov/newsroom/press/release/blumenthal-and-hawley-announce-bipartisan-framework-on-artificial-intelligence-legislation', 'snippet': 'Sep 8, 2023 ... [WASHINGTON, D.C.] – U.S. Senators Richard Blumenthal (D-CT) and Josh Hawley (R-MO), Chair and Ranking Member of the Senate Judiciary\xa0...'}, {'title': 'U.S. Artificial Intelligence Policy: Legislative and Regulatory ...', 'link': 'https://www.cov.com/en/news-and-insights/insights/2023/10/us-artificial-intelligence-policy-legislative-and-regulatory-developments', 'snippet': "Oct 20, 2023 ... Separate from Leader Schumer's effort, Senators Richard Blumenthal (D ... This proposal follows legislation Senators Blumenthal and Hawley\xa0..."}, {'title': "The Future is Here: Senate Judiciary Committee's Oversight of AI ...", 'link': 'https://www.crowell.com/en/insights/client-alerts/the-future-is-here-senate-judiciary-committees-oversight-of-ai-and-principles-for-regulation', 'snippet': 'Jul 25, 2023 ... ... AI systems. Ranking Member Josh Hawley (R-MO) gave a shorter statement, identifying his main priorities as workers, children, consumers, and\xa0...'}, {'title': 'THE PHILOSOPHY OF AI: LEARNING FROM HISTORY, SHAPING ...', 'link': 'https://www.congress.gov/118/chrg/CHRG-118shrg53996/CHRG-118shrg53996.pdf', 'snippet': 'Nov 8, 2023 ... Present: Senators Peters [presiding], Hassan, Rosen, Blumenthal,. Ossoff, Butler, Johnson, and Hawley. OPENING STATEMENT OF SENATOR PETERS1.'}, {'title': 'Hawley, Blumenthal Introduce Bipartisan Legislation to Protect ...', 'link': 'https://www.hawley.senate.gov/hawley-blumenthal-introduce-bipartisan-legislation-protect-consumers-and-deny-ai-companies-section/', 'snippet': 'Jun 14, 2023 ... Today U.S. Senators Josh Hawley (R-Mo.) ... Last week, Senator Hawley announced five guiding principles for the future of AI legislation.'}, {'title': 'CED Issues Statement on Ensuring Safe, Accessible, Credible', 'link': 'https://www.conference-board.org/press/CED-statement-safe-accessible-credible-elections', 'snippet': 'CED Issues Statement on Ensuring Safe, Accessible, Credible Elections. 2022-11-04. Dr. Lori Esposito Murray, President of the ... Explore the Impact of AI on Your Business. Members receive complimentary registration - Learn more >>'}, {'title': 'Pew Research Center | Numbers, Facts and Trends Shaping Your ...', 'link': 'https://www.pewresearch.org/', 'snippet': 'Pew Research Center is a nonpartisan, nonadvocacy fact tank that informs the public about the issues, attitudes and trends shaping the world.'}, {'title': 'Why Not A.I.? Insights from HR Teams on Worker Financial Security', 'link': 'https://www.aspendigital.org/blog/ai-for-worker-financial-security/', 'snippet': 'Oct 26, 2023 ... In recent years, headlines have been rife with horror stories about the impact of artificial intelligence (AI) on human resources (HR) work.'}, {'title': 'emerging technology - Alliance For Securing Democracy', 'link': 'https://securingdemocracy.gmfus.org/tag/emerging-technology/', 'snippet': '... artificial intelligence will impact democratic institutions and elections moving forward. ... The ASD AI Election Security Handbook. Introduction The typical\xa0...'}, {'title': 'Election Officials Under Attack', 'link': 'https://documents.ncsl.org/wwwncsl/Summit/2023/Session-Resources/Election-Officials-Under-Attack-Brennan-Center-for-Justice.pdf', 'snippet': 'Jun 16, 2021 ... associations) to improve working conditions and to better empower election officials to impact election policy. ... How AI Puts Elections at Risk.'}, {'title': 'Preventing the Harms of AI-enabled Voice Cloning | Federal Trade ...', 'link': 'https://www.ftc.gov/policy/advocacy-research/tech-at-ftc/2023/11/preventing-harms-ai-enabled-voice-cloning', 'snippet': 'Nov 16, 2023 ... ... voices in a way that is hard to detect by ear. This progress in voice cloning technology offers promise for Americans in, for example\xa0...'}, {'title': 'Federal Communications Commission FCC 24-17 Before the ...', 'link': 'https://docs.fcc.gov/public/attachments/FCC-24-17A1.pdf', 'snippet': 'Feb 8, 2024 ... understand the implications of emerging AI technologies ... Bad actors are using voice cloning - a generative AI technology that uses a recording\xa0...'}, {'title': 'Voice Cloning Technology and its Legal Implications: An IP Law ...', 'link': 'https://iplawusa.com/voice-cloning-technology-and-its-legal-implications-an-ip-law-perspective/', 'snippet': "Aug 26, 2023 ... Voice cloning technology is a cutting-edge development in the domain of artificial intelligence (AI) that involves creating a digital replica of a person's\xa0..."}, {'title': "AI Voice Cloning - and Its Misuse - Has Opened a Pandora's Box of ...", 'link': 'https://ipwatchdog.com/2023/08/09/ai-voice-cloning-misuse-opened-pandoras-box-legal-issues-heres-know/id=163859/', 'snippet': 'Aug 9, 2023 ... Voice cloning, a technology

```
that enables the replication of human voices from large language models
using artificial intelligence (AI),\xa0...'}, {'title': 'Top 5 Frequently
Asked Questions About Voice Cloning Technology', 'link': 'https://www.
respeecher.com/blog/top-5-frequently-asked-questions-about-voice-cloning-
technology', 'snippet': 'Jun 4, 2024 ... Technology Used: The complexity
and sophistication of the AI and machine learning algorithms employed can
significantly impact the cost.'}]
```

Step 6: Find the Most Relevant Results

As previously noted, Google Search provides URLs for each source, but we still need to extract the actual content from these pages. This is where the **newspaper** package comes in handy—it allows us to retrieve webpage content using the .parse() method. The following code iterates through the search results and attempts to extract the text from each linked page.

```
import newspaper

pages_content = []

for result in all_results:
  try:
    article = newspaper.Article(result["link"])
    article.download()
    article.parse()

    if len(article.text) > 0:
      pages_content.append({ "url": result["link"], "text": article.text })
  except:
    continue

print("Number of pages: ", len(pages_content))
```

Output: Number of pages: 11

Step 7: Split into Chunks

The output above indicates that only 11 pages were processed instead of the expected 15. This discrepancy can occur because the newspaper library may struggle to extract content in certain cases, such as when search results lead to PDF files or when websites impose restrictions on web scraping.

Next, it's essential to split the extracted content into smaller chunks to prevent exceeding the model's input length. The code below achieves this by segmenting the text based on either newlines or spaces, depending on the structure of the content. It ensures that each chunk contains 3000 characters with an overlap of 100 characters between consecutive chunks to maintain context.

```
from langchain.text_splitter import RecursiveCharacterTextSplitter
from langchain.docstore.document import Document

text_splitter = RecursiveCharacterTextSplitter(chunk_size=3000, chunk_
overlap=100)

docs = []
for d in pages_content:
    chunks = text_splitter.split_text(d["text"])
    for chunk in chunks:
        new_doc = Document(page_content=chunk, metadata={ "source":
        d["url"] })
        docs.append(new_doc)

print("Number of chunks: ", len(docs))
```

Output: `Number of chunks:  26`

Step 7: Create Embeddings

As shown, the docs variable now contains 26 chunks of data. The next step is to identify the most relevant chunks to use as context for the large language model. To achieve this, we leverage the OpenAIEmbeddings class, which utilizes OpenAI to convert text into a vector space that captures semantic meaning.

We then proceed to embed both the document chunks and the target sentence from the main article that we want to expand. This sentence, which was selected at the start of the lesson, is stored in the text_to_change variable. By comparing embeddings, we can retrieve the most relevant chunks to enrich the expanded content.

```
from langchain.embeddings import OpenAIEmbeddings

embeddings = OpenAIEmbeddings(model="text-embedding-ada-002")

docs_embeddings = embeddings.embed_documents([doc.page_content for doc
in docs])
query_embedding = embeddings.embed_query(text_to_change)
```

To measure the relevance of document chunks, we use the **cosine similarity** metric, which calculates the distance between high-dimensional embedding vectors. This metric helps determine how closely two points are positioned within the vector space. Since embeddings capture contextual meaning, **closer vectors indicate stronger semantic similarity**, making high-scoring documents ideal sources for expansion.

We utilize the "cosine_similarity" function from the **sklearn** library to compute the similarity between each document chunk and the target sentence. This function returns the **indices of the top three most relevant chunks**, ensuring that the model receives the most meaningful context for generating expanded content.

```
import numpy as np
from sklearn.metrics.pairwise import cosine_similarity

def get_top_k_indices(list_of_doc_vectors, query_vector, top_k):
    # convert the lists of vectors to numpy arrays
    list_of_doc_vectors = np.array(list_of_doc_vectors)
    query_vector = np.array(query_vector)

    # compute cosine similarities
    similarities = cosine_similarity(query_vector.reshape(1, -1), list_of_
    doc_vectors).flatten()

    # sort the vectors based on cosine similarity
    sorted_indices = np.argsort(similarities)[::-1]

    # retrieve the top K indices from the sorted list
    top_k_indices = sorted_indices[:top_k]

    return top_k_indices
```

```
top_k = 3
best_indexes = get_top_k_indices(docs_embeddings, query_embedding, top_k)
best_k_documents = [doc for i, doc in enumerate(docs) if i in best_indexes]
```

Step 8: Extend the Sentence

Now, we can define the **prompt** using additional information retrieved from Google Search. The template includes six input variables:

- **title**: Holds the main article's title
- **text_all**: Represents the full article being processed
- **text_to_change**: The specific section of the article that requires expansion
- **doc_1, doc_2, doc_3**: The top three most relevant Google search results, used as contextual references

The rest of the code follows the same structure as the **Google query generation process**. It defines a **HumanMessage** template, ensuring compatibility with the ChatGPT API. The model is set with a **high temperature** to promote creative output. Finally, the **LLMChain** class constructs a processing chain that integrates the model and the prompt, executing the expansion task using the `.run()` method.

```
template = """You are an exceptional copywriter and content creator.

You're reading an article with the following title:
----------------
{title}
----------------

You've just read the following piece of text from that article.
----------------
{text_all}
----------------

Inside that text, there's the following TEXT TO CONSIDER that you want to
enrich with new details.
----------------
```

```
{text_to_change}
----------------

Searching around the web, you've found this ADDITIONAL INFORMATION from
distinct articles.
----------------
{doc_1}
----------------
{doc_2}
----------------
{doc_3}
----------------

Modify the previous TEXT TO CONSIDER by enriching it with information from
the previous ADDITIONAL INFORMATION.
"""

human_message_prompt = HumanMessagePromptTemplate(
    prompt=PromptTemplate(
        template=template,
        input_variables=["text_to_change", "text_all", "title", "doc_1",
        "doc_2", "doc_3"],
    )
)
chat_prompt_template = ChatPromptTemplate.from_messages([human_message_
prompt])

chat = ChatOpenAI(model_name="gpt-4o-mini", temperature=0.9)
chain = LLMChain(llm=chat, prompt=chat_prompt_template)

response = chain.run({
    "text_to_change": text_to_change,
    "text_all": text_all,
    "title": title,
    "doc_1": best_k_documents[0].page_content,
    "doc_2": best_k_documents[1].page_content,
    "doc_3": best_k_documents[2].page_content
})
```

```
print("Text to Change: ", text_to_change)
print("Expanded Variation:", response)
```

Output: Text to Change: Senators Josh Hawley and Richard Blumenthal acknowledged AI's disruptive impact and the necessity of understanding its ramifications on elections, employment, and national security. To illustrate AI's power, Blumenthal played an audio clip generated by an AI voice cloning system trained on his past speeches.
Expanded Variation: Certainly! Here's an enriched version of the previously specified text, incorporating relevant details from the additional information:

Senators Josh Hawley (R-MO) and Richard Blumenthal (D-CT), Chair and Ranking Member of the Senate Judiciary Subcommittee on Privacy, Technology, and the Law, acknowledged AI's disruptive impact and the necessity of understanding its ramifications on elections, employment, and national security. To illustrate AI's power, Blumenthal played an audio clip generated by an AI voice cloning system trained on his past speeches, showcasing the technology's potential for misuse. In light of these concerns, both senators announced a bipartisan legislative framework aimed at establishing guardrails for artificial intelligence.............

App 6: YouTube Scriptwriting Tool

The YouTube Scriptwriting Tool is an AI-driven assistant designed to help content creators craft engaging, well-structured scripts for their videos. By leveraging GPT-powered AI, this tool streamlines the scriptwriting process, ensuring compelling storytelling, clear messaging, and audience engagement.

Step 1: Install All Required Libraries and Import Them

These dependencies are essential for building a YouTube Scriptwriting Tool with AI-powered transcription, content generation, and automation. Here's why each package is used:

- **openai**: Provides access to GPT models for generating YouTube video scripts, improving structure, and enhancing content

- **langchain**: A framework that integrates LLMs, text processing, and retrieval-based AI for better script structuring
- **google-colab**: Ensures compatibility with Google Colab, allowing the tool to run smoothly in a cloud environment
- **yt_dlp**: A powerful tool for downloading YouTube videos, enabling AI-based script generation by transcribing existing content
- **langchain-community**: Extends LangChain's capabilities with community-maintained integrations for improved AI workflows
- **openai-whisper**: A state-of-the-art AI model for speech-to-text transcription, used to convert YouTube videos into text-based scripts
- **torch**: A deep learning framework required for Whisper's AI model, enabling fast and efficient transcription

```
!pip install openai langchain google-colab yt_dlp langchain-community
!pip install -U openai-whisper torch
```

```
import openai
import os
import re
import subprocess
from langchain.chat_models import ChatOpenAI
from langchain.chains import LLMChain
from langchain.prompts import PromptTemplate
from google.colab import auth
import whisper
```

Step 2: Authenticate in Google Drive As We Use Google Colab and Insert Your OpenAI API Key

```
auth.authenticate_user()
os.environ['OPENAI_API_KEY'] = input("Enter your OpenAI API Key: ")
gpt = ChatOpenAI(temperature=0.7, model_name="gpt-4")
```

Step 3: Download Your Desired YouTube Video, Extract the Audio, and Convert It to MP3

```
### Step 2: Download YouTube Audio
def download_audio(video_url):
    video_id_match = re.search(r"(?:v=|\/)([0-9A-Za-z_-]{11}).*",
    video_url)
    if not video_id_match:
        return None

    video_id = video_id_match.group(1)
    audio_filename = f"{video_id}.mp3"

    command = f"yt-dlp -x --audio-format mp3 -o '{audio_filename}'
    {video_url}"
    os.system(command)

    return audio_filename if os.path.exists(audio_filename) else None
```

Step 4: Transcribe Audio

```
### Step 3: Transcribe Audio
def transcribe_audio(audio_filename):
    model = whisper.load_model("small")
    result = model.transcribe(audio_filename)
    return result["text"]
```

This function **transcribes audio into text** using **OpenAI's Whisper model**.

1. **Loads Whisper's "small" model** (`whisper.load_model("small")`)
2. **Transcribes the given audio file** (`model.transcribe(audio_filename)`)
3. **Returns the extracted text** (`result["text"]`)

Step 5: Generate Outline

```
### Step 4: Generate an Outline
def generate_outline(transcript_text):
    outline_prompt = PromptTemplate(
        input_variables=["transcript_text"],
        template="""
        You are a professional YouTube scriptwriter. Analyze the following
        transcribed YouTube video:
        "{transcript_text}"

        Create an engaging script outline, including an introduction, key
        sections, and a conclusion.
        """
    )

    outline_chain = LLMChain(llm=gpt, prompt=outline_prompt)
    return outline_chain.run(transcript_text)
```

This function above generates a structured outline for a YouTube script from a transcribed video.

1. Defines a prompt template (**`PromptTemplate`**) that instructs the AI to analyze the transcript and create an outline with an introduction, key sections, and a conclusion
2. Creates an AI processing chain **(`LLMChain`) using `gpt` (a GPT model)**
3. Runs the AI model to generate an engaging script outline from the given transcript

Step 6: Expand the Script

```
### Step 5: Expand Script
def expand_script(outline):
    script_prompt = PromptTemplate(
        input_variables=["outline"],
        template="""
```

```
        Given the following YouTube script outline:
        {outline}

        Expand each section into a complete, engaging script with natural
        dialogue and a strong narrative flow.
        Include timestamps and suggested visuals where relevant.
        """
    )

    script_chain = LLMChain(llm=gpt, prompt=script_prompt)
    return script_chain.run(outline)
```

This function expands a script outline into a full YouTube script using AI.

1. Defines a prompt template (PromptTemplate) that instructs the AI to convert the outline into a detailed script, ensuring natural dialogue and strong narrative flow
2. Creates an AI processing chain (LLMChain) using gpt (a GPT model)
3. Runs the AI model to generate a fully developed script, including timestamps and suggested visuals for better content structuring

Step 7: Combine All and Run the Tool

```
### Step 6: Run the Tool
if __name__ == "__main__":
    video_url = input("Enter the YouTube video URL: ")

    print("\nDownloading audio from video...\n")
    audio_filename = download_audio(video_url)
    if not audio_filename:
        print("Audio download failed! Exiting.")
    else:
        print("Audio downloaded successfully!")

        print("\nTranscribing audio...\n")
        transcript_text = transcribe_audio(audio_filename)
```

```
    print("Transcript generated successfully!\n")
    print(transcript_text)

    print("\nGenerating script outline from transcript...\n")
    outline = generate_outline(transcript_text)
    print(outline)

    input("Press Enter to generate the full script...")
    print("\nExpanding into full script...\n")
    full_script = expand_script(outline)
    print(full_script)
```

Output:

```
Enter the YouTube video URL: https://www.youtube.com/shorts/9YFT5HqL5m8

Downloading audio from video...

Audio downloaded successfully!

Transcribing audio...

Transcript generated successfully!

 While loop in Python. Firstly write out the following lines of code,
 making sure you remember the colons and the indents. Save it, then run it.
 It works.

Generating script outline from transcript...

Title: Mastering the While Loop in Python

Introduction:
- Welcoming viewers to the channel and the video
- Briefly discussing the importance of understanding Python loops,
  especially the "while loop"
- Outlining the objectives for the video

Section 1: Understanding the While Loop
- Explaining what a while loop is in the context of Python
- Discussing the use cases and benefits of using while loops
```

```
Section 2: Structuring the While Loop
- Explaining the syntax of the while loop, emphasizing the importance of
  colons and indents
- Showing on screen an example of the structure of a basic while loop

Section 3: Writing the Code
- Taking viewers through the process of writing a simple while loop code
- Highlighting key points such as the use of colons and indents, how to
  structure the loop, and what each line of code does

Section 4: Saving and Running the Code
- Demonstrating how to save and run the code
- Discussing potential errors that could occur and how to troubleshoot

Conclusion:
- Recapping the importance and structure of while loops in Python
- Encouraging viewers to practice writing their own while loops
- Reminding viewers to like, share, and subscribe for more Python tutorials
- Teasing the topic of the next video and bidding viewers farewell until
  next time.
Press Enter to generate the full script...

Expanding into full script...

Title: Mastering the While Loop in Python

[Introduction 00:00]

(Visual: Channel logo animation)

HOST: "Hey there coding enthusiasts, welcome back to our channel, your
trusted guide to everything Python! We all know how crucial loops are
in Python, don't we? And today, we're diving deep into the fascinating
world of 'While Loops' in Python. We'll be exploring what they are, how to
structure them, and finally, we'll write some code together. So, let's get
started!"

(Visual: Text Animation - "Mastering the While Loop in Python")
```

[Section 1: Understanding the While Loop 00:30]

(Visual: Video Animation - "While Loop Concept")

HOST: "So what exactly is a while loop? In Python, a while loop is used for iterative tasks, which simply means, it helps you execute the same code over and over again until a certain condition is met. It's like telling your computer, 'Hey, keep doing this task while this condition is true!'. And the benefits? It's a massive time-saver and a powerful tool for handling repetitive tasks."

[Section 2: Structuring the While Loop 01:15]

(Visual: Screen recording - Python IDE with blank code file)

HOST: "Now, let's talk about how we structure a while loop in Python. The syntax is straightforward. We start with the keyword 'while', followed by the condition, and then a colon. The code you want to repeat goes underneath, indented for clarity. Let's look at a basic example."

(Visual: Coding example on Python IDE)

[Section 3: Writing the Code 02:30]

(Visual: Screen Recording - Python IDE with code example)

HOST: "Let's write a simple while loop code together, shall we? Remember, our indents and colons are key here. We'll structure our loop, line by line, and I'll explain each part as we go along."

(Visual: Host typing and explaining the code)

[Section 4: Saving and Running the Code 04:50]

(Visual: Screen Recording - Python IDE)

HOST: "Once we've written our code, it's time to save and run it. But remember, errors can occur. Maybe we've missed a colon or misstructured our loop. Don't worry, I'll show you common errors and how to troubleshoot them."

(Visual: Demonstration of saving, running, and troubleshooting the code)

[Conclusion 06:40]

```
(Visual: Host on screen)

HOST: "And that, my friends, is the while loop in Python! Remember,
practice is key, so try writing your own while loops. Don't forget to hit
the like button if you found this tutorial helpful, and share it with your
fellow coders. Subscribe for more Python tutorials, and stay tuned for our
next video where we'll be delving into another exciting Python topic. Until
then, keep coding!"

(Visual: End screen with like, share, and subscribe animation)
```

App 7: Email Generator

The AI Email Generator is a powerful tool designed to automate and enhance email writing using AI. By leveraging GPT-powered language models, this tool helps users craft professional, personalized, and context-aware emails in seconds.

Key Features

- **Automated Email Drafting**: Generate emails based on prompts or key points.
- **Personalization**: Adjust tone, style, and recipient details for a tailored approach.
- **Quick Edits and Refinements**: Modify content instantly with AI suggestions.
- **Template-Based Generation**: Create emails for business, customer support, marketing, and more.
- **Grammar and Tone Enhancement**: Ensure clarity, professionalism, and engagement.

Ideal for professionals, businesses, and individuals, the AI Email Generator streamlines communication, saves time, and improves email effectiveness with AI-driven precision.

Step 1: Install All Required Libraries and Import Them

```
!pip install langchain openai langchain_community

from langchain_openai import ChatOpenAI
from langchain.prompts import PromptTemplate
from langchain.chains import LLMChain
import os
```

These dependencies are essential for building an AI-powered Email Generator using LangChain and OpenAI. Here's why each package is needed:

- `langchain`: The core framework for integrating LLMs (like GPT-4) to generate, refine, and personalize email content
- `openai`: Provides access to GPT-powered AI for drafting professional, context-aware emails
- `langchain_community`: Enhances LangChain with community-supported integrations for better performance and extended capabilities

This setup enables smart, AI-driven email generation, making the process faster, more efficient, and highly personalized.

Step 2: Generate Response with OpenAI

```
def generate_email_response(api_key, original_email, sender_name,
recipient_name, response_tone="professional"):

    os.environ["OPENAI_API_KEY"] = api_key

    template = PromptTemplate(
        input_variables=["original_email", "sender_name", "recipient_name",
        "response_tone"],
        template="""
        Read the following email from {sender_name} and generate a well-
        structured, contextually relevant response for {recipient_name}.
        Ensure the tone of the response is {response_tone} and
        appropriately addresses the content of the original email.
```

```
        Original Email:
        {original_email}

        Keep the response concise yet informative, maintaining politeness
        and clarity.
        """,
    )

    llm = OpenAI(model="gpt-3.5-turbo")
    chain = LLMChain(llm=llm, prompt=template)

    response_email = chain.run({
        "original_email": original_email,
        "sender_name": sender_name,
        "recipient_name": recipient_name,
        "response_tone": response_tone
    })

    return response_email
```

In this code, the `generate_email_response()` function takes an API key, an original email, the sender and recipient names, and an optional response tone (defaulting to "professional").

It first sets the OpenAI API key as an environment variable (`os.environ["OPENAI_API_KEY"] = api_key`) to authenticate requests to OpenAI's API.

A `PromptTemplate` is then defined, guiding the AI to read the original email and generate a contextually relevant, well-structured response. The AI

- Adapts the tone (e.g., professional, friendly)
- Addresses the recipient appropriately
- Keeps the response concise, polite, and informative

An LLM model (`OpenAI()`) is initialized, and an LLMChain (`chain`) is created to process the prompt dynamically.

The function executes the chain with the given email details, generating an AI-written email response, which is then returned.

This setup saves time, enhances professionalism, and ensures clarity, making it useful for customer support, business communication, and automated email responses.

Step 3: Combine All Together and Generate Email

```
if __name__ == "__main__":
    api_key = input("Enter your OpenAI API key: ")
    original_email = input("Enter the original email content: ")
    sender_name = input("Enter the sender's name: ")
    recipient_name = input("Enter the recipient's name: ")
    response_tone = input("Enter the response tone (e.g., professional,
    friendly, casual): ")

    response = generate_email_response(api_key, original_email, sender_
    name, recipient_name, response_tone)

    print("\nGenerated Email Response:\n")
    print(response)
```

Note Don't forget to generate your OpenAI API key.

Output:

```
Enter your OpenAI API key:
Enter the original email content: Let's have a meeting together?
Enter the sender's name: Anthony
Enter the recipient's name: James
Enter the response tone (e.g., professional, friendly, casual):
Professional
<ipython-input-5-279420fd8784>:30: LangChainDeprecationWarning: The
class `OpenAI` was deprecated in LangChain 0.0.10 and will be removed in
1.0. An updated version of the class exists in the :class:`~langchain-
openai package and should be used instead. To use it run `pip install -U
:class:`~langchain-openai` and import as `from :class:`~langchain_openai
import OpenAI``.
  llm = OpenAI()
<ipython-input-5-279420fd8784>:31: LangChainDeprecationWarning: The class
`LLMChain` was deprecated in LangChain 0.1.17 and will be removed in 1.0.
Use :meth:`~RunnableSequence, e.g., `prompt | llm`` instead.
```

```
  chain = LLMChain(llm=llm, prompt=template)
<ipython-input-5-279420fd8784>:33: LangChainDeprecationWarning: The method
`Chain.run` was deprecated in langchain 0.1.0 and will be removed in 1.0.
Use :meth:`~invoke` instead.
  response_email = chain.run({
Generated Email Response:

Dear Anthony,

Thank you for reaching out to me about having a meeting together. I am
always open to discussing and collaborating on any important matters.

Could you please provide more details about the meeting? This will help
me prepare and make the most of our time together. Additionally, please
suggest a few dates and times that work for you so we can schedule the
meeting accordingly.

I look forward to meeting with you and discussing further.

Best regards,
James
```

App 8: CSV Data Analysis App

The CSV Data Analysis App is a powerful tool designed to help users efficiently analyze, visualize, and extract insights from structured datasets. By leveraging AI, data processing libraries, and interactive visualizations, this app makes it easy to explore large CSV files, perform statistical analysis, and generate meaningful reports.

Step 1: Install All Required Libraries and Import Them

```
!pip install pandas langchain openai matplotlib seaborn langchain_community
langchain_experimental

import pandas as pd
import langchain
from langchain.llms import OpenAI
from langchain_experimental.agents import create_pandas_dataframe_agent
```

```
import matplotlib.pyplot as plt
import seaborn as sns
import os
```

These dependencies enable a CSV Data Analysis App by integrating AI, data processing, and visualization:

- **pandas**: Loads and manipulates CSV files.
- **langchain and openai**: Uses GPT-4 for AI-powered insights and queries
- **matplotlib and seaborn**: Creates professional data visualizations
- **langchain_community and langchain_experimental**: Enhances AI integration with modern tools

This setup allows users to analyze, visualize, and gain AI-driven insights from CSV datasets, making data exploration faster and smarter.

Step 2: Generate and Add Your OpenAI API Key

```
# Set API keys (Use environment variables for security)
import os
os.environ["OPENAI_API_KEY"] = <Your API Key>
```

Step 3: Load Your CSV File

This code below loads, previews, and analyzes a CSV file using pandas.

The `load_csv(file_path)` function takes a file path as input and loads the CSV file into a pandas DataFrame using `pd.read_csv()`.

It then prompts the user to enter the CSV file path, loads the data into `df`, and displays key insights:

1. Data Preview: Prints the first five rows of the dataset using `df.head()`, providing a quick look at the data structure
2. Basic Statistics: Prints summary statistics with `df.describe()`, showing key metrics like mean, min, max, and standard deviation for numerical columns

```
# Function to load CSV
def load_csv(file_path):
    return pd.read_csv(file_path)

# Load CSV
file_path = input("Enter the path to your CSV file: ")
df = load_csv(file_path)

# Display data preview
print("\n### Data Preview")
print(df.head())

# Display basic statistics
print("\n### Basic Statistics")
print(df.describe())
```

Note To get the path of a file in Google Colab, upload the file, right-click with your cursor, and select "Copy path."

Output:

```
Enter the path to your CSV file: /content/langchain_broad-match_
us_2025-02-19.csv

### Data Preview
                   Keyword         Intent  Volume  \
0             langchain js   Navigational    1000
1         langchain openai  Informational    1000
2          langchain tools  Informational    1000
3       langchain tutorial   Navigational    1000
4  langchain_community.llms  Informational    1000

                                               Trend  Keyword Difficulty  \
0  1.00,0.38,1.00,0.68,0.46,0.68,0.38,0.52,0.52,0...                  44
1  0.62,1.00,0.81,0.62,0.36,0.62,0.45,0.55,0.45,0...                  35
2  0.37,0.52,0.08,1.00,0.68,0.52,0.68,0.52,0.52,0...                  41
```

```
3  0.44,0.44,0.81,0.81,0.62,0.62,0.55,0.55,1.00,0...                  54
4  0.13,0.36,1.00,0.54,0.36,0.02,0.04,0.03,0.00,0...                  18

   CPC (USD)  Competitive Density  \
0       0.00                 0.01
1       3.63                 0.01
2       2.48                 0.00
3       2.98                 0.11
4       0.00                 0.00

                                        SERP Features  Number of Results
0  Sitelinks, Video, People also ask, Related sea...            6260000
1           Video, People also ask, Related searches           11600000
2                 Sitelinks, Video, Related searches           16400000
3  Sitelinks, Video, People also ask, Related sea...            8510000
4                                  Image pack, Video                 42

### Basic Statistics
           Volume  Keyword Difficulty   CPC (USD)  Competitive Density  \
count  284.000000          284.000000  284.000000           284.000000
mean   288.204225           28.419014    0.878275             0.019014
std    200.324685           12.485063    2.212721             0.069697
min    110.000000            0.000000    0.000000             0.000000
25%    140.000000           20.000000    0.000000             0.000000
50%    210.000000           27.000000    0.000000             0.000000
75%    320.000000           36.000000    0.000000             0.010000
max   1000.000000           83.000000   17.340000             0.830000

       Number of Results
count       2.840000e+02
mean        3.444161e+06
std         5.098099e+06
min         0.000000e+00
25%         9.700000e+01
50%         5.330000e+05
75%         5.745000e+06
max         3.630000e+07
```

Step 4: Create a LangChain Agent

```
# LangChain Agent for querying data
OPENAI_API_KEY = os.environ["OPENAI_API_KEY"]
# LangChain Agent for querying data
if OPENAI_API_KEY:
    llm = OpenAI(temperature=0, openai_api_key=OPENAI_API_KEY)
    agent = create_pandas_dataframe_agent(llm, df, verbose=True, allow_
    dangerous_code=True)

    query = input("\nAsk a question about the data: ")
    if query:
        print("\nAnalyzing...")
        response = agent.run(query)
        print("\n**Response:**", response)
else:
    print("\nWarning: OpenAI API Key not found. Please set it as an
    environment variable.")
```

The code above sets up a LangChain agent to interact with a CSV dataset using GPT-powered AI queries.

First, it retrieves the OpenAI API key from the environment (`os.environ["OPENAI_API_KEY"]`). If the key exists, it initializes an LLM instance (`OpenAI`) with `temperature=0` for deterministic responses.

It then creates a Pandas DataFrame agent using `create_pandas_dataframe_agent(llm, df, verbose=True, allow_dangerous_code=True)`. This agent allows users to ask natural language questions about the dataset, and the AI will analyze and generate insights based on the data.

The program then prompts the user for a query. If a question is provided, the agent processes the request, runs the query on the DataFrame, and returns an AI-generated response.

If the API key is missing, it prints a warning message, instructing the user to set up the key.

This setup enables AI-powered data analysis, allowing users to interact with CSV datasets using natural language instead of manual coding.

Output:

```
Ask a question about the data: What's the data about?

Analyzing...

> Entering new AgentExecutor chain...
<ipython-input-16-9aca66b979d8>:11: LangChainDeprecationWarning: The method
`Chain.run` was deprecated in langchain 0.1.0 and will be removed in 1.0.
Use :meth:`~invoke` instead.
  response = agent.run(query)
Thought: The data is about keywords and their corresponding attributes.
Action: python_repl_ast
Action Input: df.info()<class 'pandas.core.frame.DataFrame'>
RangeIndex: 284 entries, 0 to 283
Data columns (total 9 columns):
 #   Column               Non-Null Count  Dtype
---  ------               --------------  -----
 0   Keyword              284 non-null    object
 1   Intent               284 non-null    object
 2   Volume               284 non-null    int64
 3   Trend                284 non-null    object
 4   Keyword Difficulty   284 non-null    int64
 5   CPC (USD)            284 non-null    float64
 6   Competitive Density  284 non-null    float64
 7   SERP Features        284 non-null    object
 8   Number of Results    284 non-null    int64
dtypes: float64(2), int64(3), object(4)
memory usage: 20.1+ KB
 The data has 284 rows and 9 columns.
Action: python_repl_ast
Action Input: df.shape(284, 9)I now know the final answer
Final Answer: The data has 284 rows and 9 columns. It contains information
about keywords, their intent, volume, trend, keyword difficulty, CPC,
competitive density, SERP features, and number of results.

> Finished chain.
```

App 9: Knowledge Base Voice Assistant

The Knowledge Base Voice Assistant is an AI-driven system that enables users to interact with a knowledge base using natural voice commands. By integrating speech recognition, large language models (LLMs), and vector search, this assistant allows for seamless and intelligent access to vast amounts of information.

Designed for businesses, research teams, and customer support, this voice-enabled assistant can retrieve answers, summarize documents, and provide real-time insights from structured and unstructured data sources. By leveraging LangChain, OpenAI's GPT models, and vector databases, the assistant delivers accurate and context-aware responses in a conversational format.

Step 1: Install the Required Libraries and Import Them

```
# Install dependencies
!pip install SpeechRecognition gtts langchain faiss-cpu openai
```

```
import speech_recognition as sr
from gtts import gTTS
import os
import IPython.display as ipd
from langchain.vectorstores import FAISS
from langchain.embeddings.openai import OpenAIEmbeddings
from langchain.chat_models import ChatOpenAI
from langchain.chains import RetrievalQA
from langchain.document_loaders import WebBaseLoader
from langchain.text_splitter import CharacterTextSplitter
from langchain.vectorstores import FAISS
from google.colab import files
```

These dependencies enable a **voice-controlled AI assistant** by integrating **speech recognition, retrieval, and AI-powered responses**:

- **SpeechRecognition**: Converts speech to text for voice input
- **gTTS**: Converts AI-generated text to speech for voice output
- **LangChain**: Manages LLM interactions and knowledge retrieval

- **FAISS**: Enables fast, semantic search in the knowledge base
- **OpenAI**: Uses GPT-4 to generate intelligent responses

Together, these tools allow users to **speak queries, retrieve relevant information, and hear AI-generated answers**, making knowledge access seamless and intuitive.

Step 2: Generate and Add Your OpenAI API Key

```
# Set API keys (Use environment variables for security)
import os
os.environ["OPENAI_API_KEY"] = <Your API Key>
```

Step 3: Develop Voice Interaction

```
def speak(text):
    """Convert text to speech using gTTS and play it."""
    tts = gTTS(text=text, lang='en')
    tts.save("response.mp3")
    ipd.display(ipd.Audio("response.mp3"))

def listen():
    """Process uploaded audio file and convert to text."""
    recognizer = sr.Recognizer()
    print("Please upload an audio file (wav format).")
    uploaded = files.upload()

    for filename in uploaded.keys():
        with sr.AudioFile(filename) as source:
            audio = recognizer.record(source)
        try:
            return recognizer.recognize_google(audio)
        except sr.UnknownValueError:
            return "Sorry, I could not understand."
        except sr.RequestError:
            return "Could not request results."
```

This code enables voice interaction for an AI assistant by handling both text-to-speech (TTS) output and speech-to-text (STT) input using gTTS and SpeechRecognition.

The `speak(text)` function takes a text input, converts it into speech using gTTS (Google Text-to-Speech), and saves the generated audio as `"response.mp3"`. It then plays the audio using IPython's audio player (`ipd.Audio`), allowing users to hear the assistant's response.

The `listen()` function processes an uploaded audio file (in `.wav` format) and converts speech into text. It uses SpeechRecognition's Recognizer to handle transcription. **The user is prompted to upload an audio file, which is then processed:**

1. Loads the uploaded file
2. Extracts audio data using `sr.AudioFile()`
3. Recognizes speech using Google's speech-to-text API (`recognize_google`)
4. Returns the transcribed text or an error message if speech is unclear (`UnknownValueError`) or if the API request fails (`RequestError`)

This setup allows the assistant to listen to user queries, process them as text, and respond with AI-generated speech, enabling a full voice-based knowledge assistant experience.

Note Since we develop this app in Google Colab, this platform doesn't provide us with microphone access, so all audio interaction has to be recorded as **.wav files and uploaded to Google Colab.**

Step 4: Load Knowledge Base from the Web and Create the QA Chain

```
# Load knowledge base from the web
def load_knowledge_base():
    """Load and process online resources for retrieval."""
    urls = [
        "https://en.wikipedia.org/wiki/Artificial_intelligence",
```

```
        "https://en.wikipedia.org/wiki/Natural_language_processing"
    ]

    loader = WebBaseLoader(urls)
    documents = loader.load()
    text_splitter = CharacterTextSplitter(chunk_size=1000, chunk_
    overlap=200)
    texts = text_splitter.split_documents(documents)

    embeddings = OpenAIEmbeddings()
    vectorstore = FAISS.from_documents(texts, embeddings)
    return vectorstore

def create_qa_chain(vectorstore):
    """Set up LangChain's RetrievalQA model."""
    llm = ChatOpenAI()
    retriever = vectorstore.as_retriever()
    return RetrievalQA.from_chain_type(llm=llm, retriever=retriever)
```

The code above fetches knowledge from the Web, processes it into a retrievable format, and sets up an AI-powered Q&A system using LangChain.

The `load_knowledge_base()` function

1. Defines a list of URLs containing knowledge (Wikipedia pages on AI and NLP)
2. Uses `WebBaseLoader` to fetch and extract the content from these web pages
3. Splits the extracted text into chunks of 1000 characters, ensuring 200-character overlap for better context retention using `CharacterTextSplitter`
4. Converts these text chunks into vector embeddings using `OpenAIEmbeddings`, allowing semantic search
5. Stores the processed embeddings in a FAISS vector database, which enables efficient retrieval of relevant knowledge

The `create_qa_chain(vectorstore)` function

1. Initializes an LLM-powered chatbot using `ChatOpenAI()`
2. Converts the FAISS vector store into a retriever, allowing the AI to find relevant information from stored knowledge
3. Creates a retrieval-based question-answering (QA) system using `RetrievalQA.from_chain_type()`, enabling users to ask natural language questions and get context-aware answers

This setup allows an AI assistant to retrieve and answer questions based on web-sourced knowledge, making it useful for automated research assistants, chatbots, and real-time information retrieval systems.

Step 5: Combine Them All Together

```
def main():
    """Main loop for voice interaction."""
    vectorstore = load_knowledge_base()
    qa_chain = create_qa_chain(vectorstore)

    speak("Hello! Please upload an audio file with your query.")
    while True:
        query = listen()
        if query.lower() in ["exit", "quit", "stop"]:
            speak("Goodbye!")
            break

        print(f"User: {query}")
        response = qa_chain.run(query)
        print(f"Assistant: {response}")
        speak(response)
```

This code creates a voice-interactive AI assistant that retrieves information from a web-based knowledge base and responds using speech.

The `main()` function

1. Loads the knowledge base by calling `load_knowledge_base()`, which fetches and processes online content into a retrievable format
2. Creates a Q&A system using `create_qa_chain(vectorstore)`, allowing AI-driven responses based on stored knowledge
3. Welcomes the user with speech using `speak("Hello! Please upload an audio file with your query.")`, prompting them to submit a voice query
4. Enters a loop where
   - It listens for user input via `listen()`, which converts speech into text
   - If the user says `"exit"`, `"quit"`, or `"stop"`, the assistant ends the conversation with a goodbye message
   - Otherwise, it retrieves and generates an AI-powered response using `qa_chain.run(query)`, prints it, and speaks the response aloud using `speak(response)`

Output—if all works correctly:

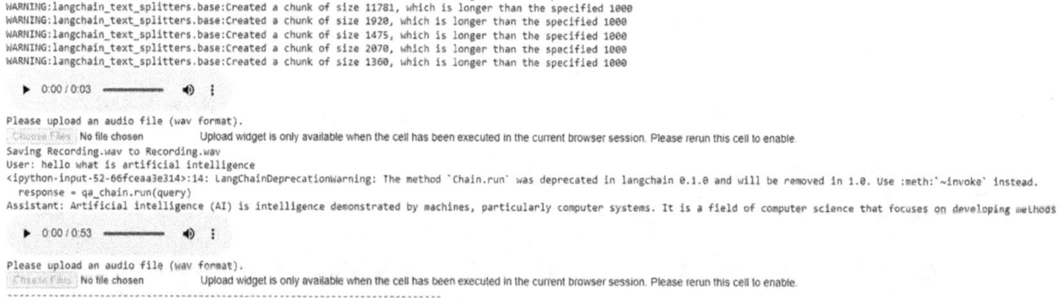

Figure 3-1. *Voice Assistant Output*

App 10: Analyzing Codebase with LangChain

The Analyzing Codebase with LangChain app is an AI-powered tool designed to help developers, engineers, and teams efficiently explore and understand complex code bases. By leveraging LangChain's advanced language processing capabilities, the app can extract insights, answer questions, and provide recommendations based on the structure and logic of a given code repository.

Using large language models (LLMs) like GPT-4, along with vector search and semantic retrieval, this app enables users to quickly navigate source code, identify dependencies, summarize functions, and even detect potential issues—all without manually scanning through thousands of lines of code.

Step 1: Install All Required Libraries

```
!pip install langchain openai chromadb tiktoken
!pip install -U langchain-community
!pip install unstructured
```

These installation commands ensure that all necessary dependencies are available for building an AI-powered code base analysis tool using LangChain. Here's why each package is needed:

- **`langchain`**: The core framework that enables interaction with large language models (LLMs), vector databases, and advanced AI tools for processing and analyzing code.
- **`openai`**: Provides access to OpenAI's models (like GPT-4), which can generate insights, summarize code, and answer questions intelligently.
- **`chromadb`**: A vector database used for efficient storage and retrieval of embeddings. This is crucial for semantic search, allowing the AI to find relevant code snippets quickly.
- **`tiktoken`**: A tokenizer for OpenAI models that helps efficiently count and manage tokens, ensuring that the AI processes code efficiently while staying within model constraints.

- **langchain-community**: The updated package containing community-supported integrations for third-party tools like ChromaDB, OpenAI, FAISS, and more. Keeping this updated ensures compatibility with the latest LangChain features.
- **unstructured**: A powerful library for extracting and processing text from complex files and documents, including code files, PDFs, and markdowns. This helps in parsing, cleaning, and structuring raw code data before embedding it into a vector database.
- As an alternative, you can use Docling. It is an innovative document processing tool developed by xAI, designed to streamline the extraction and analysis of information from various file formats like PDFs, images, and text documents. It leverages advanced AI techniques to enable users to quickly interpret and interact with complex documents, making it a valuable asset for research, data analysis, and knowledge management.

Step 2: Generate and Add Your OpenAI API Key

```
# Set API keys (Use environment variables for security)
import os
os.environ["OPENAI_API_KEY"] = <Your API Key>
```

Step 3: Upload and Load the Files

The next lines of code **scan a directory** for Python (.py) files, **reads their content**, and **stores them in a structured format** for further processing, such as embedding or AI-powered analysis.

The load_code_files function does the following:

1. Uses the glob module to **find all Python files** (.py) within the specified directory ("./my_codebase") and its subdirectories (recursive=True)
2. Initializes an empty list called documents to store the extracted code

3. Iterates over each Python file found:
   - Opens the file in **read mode** with UTF-8 encoding
   - Reads the entire content of the file
   - Appends a dictionary containing the file's **path** and **content** to the `documents` list
4. Returns the `documents` list, which now contains the **path and source code** of each Python file

The **example usage** calls `load_code_files()`, storing the result in `documents`. It then prints the total number of Python files loaded.

```
import glob

def load_code_files(directory="./my_codebase"):
    code_files = glob.glob(f"{directory}/**/*.py", recursive=True)
    documents = []
    for file_path in code_files:
        with open(file_path, "r", encoding="utf-8") as file:
            documents.append({"path": file_path, "content": file.read()})
    return documents

# Example Usage
documents = load_code_files()
print(f"Loaded {len(documents)} code files.")
```

Step 4: Create and Store Code Embeddings

This code **automates the process of loading, processing, and embedding a code base** for efficient search and retrieval using **LangChain and OpenAI embeddings**.

First, it uses **`DirectoryLoader`** to scan the `./my_codebase` directory (feel free to change it according to your needs) for all Python files (`**/*.py`) (you can look for the file extension according to your programming language). It loads these files into memory as **documents**, displaying a progress indicator during the process.

Next, the **`RecursiveCharacterTextSplitter`** is used to **split the loaded code into smaller chunks** of 500 characters, with a 50-character overlap between chunks. This ensures that when the AI retrieves and processes code, it maintains context across split sections.

After splitting the code, **embeddings** are generated using OpenAI's **OpenAIEmbeddings**, which converts each chunk into a **vector representation**. These embeddings allow for **semantic search**, meaning the AI can find relevant code snippets based on meaning rather than just exact keyword matches.

Finally, these embeddings are stored in a **Chroma vector database** using `Chroma.from_documents(docs, embedding_model)`, making the code searchable and retrievable based on AI-powered similarity searches.

This setup enables **AI-powered code search, understanding, and analysis**, making it useful for **automated documentation, intelligent code retrieval, and AI-assisted debugging**.

```
from langchain.document_loaders import DirectoryLoader
from langchain.text_splitter import RecursiveCharacterTextSplitter
from langchain.embeddings.openai import OpenAIEmbeddings
from langchain.vectorstores import Chroma

# Load code files
loader = DirectoryLoader("./my_codebase", glob="**/*.py", show_
progress=True)
documents = loader.load()

# Split code into chunks
splitter = RecursiveCharacterTextSplitter(chunk_size=500, chunk_overlap=50)
docs = splitter.split_documents(documents)

# Generate embeddings
embedding_model = OpenAIEmbeddings()
vectorstore = Chroma.from_documents(docs, embedding_model)

print("Code embeddings stored successfully!")
```

Step 5: Create Retriever and Retrieval Chain

```
from langchain.chains import create_retrieval_chain
from langchain.chat_models import ChatOpenAI
from langchain.schema.runnable import RunnablePassthrough
from langchain.prompts import ChatPromptTemplate
```

```
# Setup LLM
llm = ChatOpenAI(model="gpt-4", temperature=0)

# Create retriever
retriever = vectorstore.as_retriever(search_kwargs={"k": 5})

# Define the prompt template
prompt = ChatPromptTemplate.from_template(
    "You are an AI code assistant. Explain the following query based on the
    code context:\n\nQuery: {input}\n\nContext: {context}"
)

# Create retrieval chain
qa_chain = create_retrieval_chain(retriever, prompt | llm |
RunnablePassthrough())

# Function to Query the Codebase
def query_codebase(query):
    result = qa_chain.invoke({"input": query})  # Run the query
    print("\n Full Response:", result)  # Debug: print full output
    return result['answer']

# Example Usage
query = "How does the class Tomorrow work?"
response = query_codebase(query)
print("\n Search Result:\n", response)
```

This code **sets up an AI-powered code retrieval and explanation system** using **LangChain**, allowing users to ask questions about a code base and receive intelligent, context-aware responses.

First, it initializes a **GPT-4 model** using `ChatOpenAI` with `temperature=0`, ensuring **deterministic responses** for accurate code explanations.

Next, a **retriever** is created from the `vectorstore`, configured to return the **top five most relevant code snippets** (`k=5`) when queried. This ensures that only the most relevant parts of the code base are retrieved for explanation.

A **prompt template** is then defined using `ChatPromptTemplate.from_template()`, instructing the AI to act as a **code assistant**. It dynamically inserts

- **The user query** (`{input}`)
- **The retrieved code context** (`{context}`)

This ensures that responses are directly based on the actual code base.

The **retrieval chain** (`qa_chain`) is then created using `create_retrieval_chain()`. It follows a structured pipeline:

1. **Retrieve relevant code snippets** (`retriever`)
2. **Format the query and retrieved context into the prompt** (`prompt`)
3. **Generate an AI-powered explanation using GPT-4** (`llm`)
4. **Pass through the final response** (`RunnablePassthrough()`)

The `query_codebase` function allows users to input a **natural language question about the code base**. It runs the `qa_chain`, processes the query, and returns an AI-generated response. For debugging, it also prints the **full response object**.

Finally, an **example query** is run:

- The user asks **"How does the class Tomorrow work?" (replace it with your own query)**.
- The system searches the **vectorized code base** for relevant code snippets.
- GPT-4 generates an explanation based on the retrieved context.
- The response is printed, providing a clear AI-generated answer about the code.

This setup enables **AI-powered code analysis**, making it useful for **developer assistance, code documentation, debugging, and understanding large code bases**.

Output:

```
Search Result:
 content="The `Tomorrow` class is designed to handle asynchronous tasks
in Python. It uses the `concurrent.futures.ThreadPoolExecutor` to manage
a pool of worker threads that can execute tasks in parallel.\n\nHere's a
```

```
breakdown of how the `Tomorrow` class works:\n\n- The `__init__` method
initializes an instance of the `Tomorrow` class with a `future` object
and a `timeout` value. The `future` object represents a computation or
I/O bound task that hasn't completed yet.\n\n- The `__getattr__` method
waits for the result of the `future` object and then returns the attribute
with the specified name from the result.\n\n- The `result` property is a
shortcut for getting the result of the `future` object.\n\n- The `__iter__`
method allows an instance of the `Tomorrow` class to be iterable. It waits
for the result of the `future` object and then returns an iterator for the
result.\n\n- The `_wait` method waits for the result of the `future` object
with the specified timeout and then returns the result.\n\nThe `async_`
function is a decorator that makes a function run asynchronously. It takes
a number of threads `n`, a `base_type` which should be a type of executor,
and an optional `timeout`. It returns a `Tomorrow` object that represents
the asynchronous execution of the function.\n\nThe `threads` function is
a shortcut for using the `async_` decorator with `ThreadPoolExecutor`
as the `base_type`. It takes a number of threads `n` and an optional
`timeout`, and returns a decorator that makes a function run asynchronously
using a thread pool." additional_kwargs={} response_metadata={'token_
usage': {'completion_tokens': 338, 'prompt_tokens': 887, 'total_tokens':
1225, 'completion_tokens_details': {'accepted_prediction_tokens': 0,
'audio_tokens': 0, 'reasoning_tokens': 0, 'rejected_prediction_tokens':
0}, 'prompt_tokens_details': {'audio_tokens': 0, 'cached_tokens': 0}},
'model_name': 'gpt-4', 'system_fingerprint': None, 'finish_reason': 'stop',
'logprobs': None} id='run-8368c60d-3758-41b5-ad9f-6f000b64c48d-0'
```

App 11: Recommender System with LangChain

A **Recommender System with LangChain** is an AI-powered system that suggests relevant content, products, or information by leveraging **LangChain**'s capabilities in natural language processing, vector databases, and large language models (LLMs). It integrates **retrieval-augmented generation (RAG)** techniques, semantic search, and embeddings to provide intelligent and personalized recommendations.

How It Works

1. **Data Ingestion and Processing**: The system processes and structures input data, which could be product descriptions, research papers, articles, or user preferences.
2. **Text Embedding and Vector Storage**: Text data is converted into embeddings using models like **`OpenAIEmbeddings`**. The embeddings are stored in a **vector database** like FAISS, Pinecone, or ChromaDB.
3. **Retrieval Based on Similarity**: When a user queries the system, their input is also converted into an embedding and compared against stored embeddings to find the most relevant matches.
4. **LLM-Enhanced Recommendations**: A language model (such as GPT-4) can refine and explain the recommendations by generating context-aware suggestions based on retrieved results.
5. **Personalization and Context Memory**: By integrating **memory mechanisms** like **`ConversationBufferMemory()`**, the system can refine recommendations based on user preferences and past interactions.

Step 1: Install and Import the Required Libraries

```
!pip install langchain openai==0.28 faiss-cpu tiktoken
!pip install -U langchain-community
```

These installation commands ensure that all necessary dependencies are available for building a **LangChain-based AI system**, such as a **PDF chatbot or a recommender system**. Here's why each package is used:

- **`langchain`**: The core framework for integrating large language models (LLMs), vector databases, and retrieval-based AI applications.
- **`openai==0.28`**: Installs version 0.28 of the OpenAI Python package, which allows interaction with GPT-4, GPT-3.5, and embeddings models. Using a specific version ensures compatibility with LangChain.

- **faiss-cpu**: A vector database optimized for fast similarity search, used to store and retrieve text embeddings efficiently.
- **tiktoken**: A tokenizer for OpenAI models that helps optimize token usage and ensures the chatbot stays within token limits.
- **langchain-community**: The updated package for community-supported integrations, replacing older LangChain modules for better maintenance and compatibility with third-party tools like FAISS, Pinecone, and OpenAI.

Step 2: Generate and Add Your OpenAI API Key and Then Import All Libraries Required

```
from langchain.schema import Document
from langchain.vectorstores import FAISS
from langchain.chains import RetrievalQA
from langchain.memory import ConversationBufferMemory
from langchain.embeddings import OpenAIEmbeddings
from langchain.chat_models import ChatOpenAI
from langchain.tools import Tool
from langchain.agents import initialize_agent

# Set API keys (Use environment variables for security)
import os
os.environ["OPENAI_API_KEY"] = <Your API Key>
```

Step 3: Load Up Some Sample Data

```
# Sample data (list of items to recommend)
data = [
    "The Lord of the Rings - A fantasy novel by J.R.R. Tolkien.",
    "Harry Potter - A young wizard's journey by J.K. Rowling.",
    "The Matrix - A sci-fi movie about a simulated reality.",
    "Inception - A movie about dreams within dreams.",
    "The Witcher - A fantasy book and TV series about a monster hunter.",
    "Game of Thrones - A TV series based on A Song of Ice and Fire.",
```

```
    "Interstellar - A sci-fi movie about space exploration.",
    "Dune - A sci-fi novel by Frank Herbert about interstellar politics.",
    "Blade Runner - A dystopian sci-fi movie exploring artificial
    intelligence.",
    "Neuromancer - A cyberpunk novel by William Gibson about hackers
    and AI."
]
```

Step 4: Convert Data into LangChain Document Format and Generate Embeddings

```
# Convert data into LangChain Document format
documents = [Document(page_content=item) for item in data]

# Generate Embeddings
embeddings = OpenAIEmbeddings()
vector_store = FAISS.from_documents(documents, embeddings)
```

The code above processes raw text data, converts it into LangChain's **Document** format, and then generates vector embeddings to store in a **FAISS database** for efficient retrieval.

The first line transforms each item in the `data` list into a **LangChain Document** object, which is a standardized format for handling text in LangChain-based applications. This is necessary because LangChain's retrieval mechanisms expect data to be in this structured format.

Next, an instance of **OpenAIEmbeddings** is created. This model converts text into **numerical vector representations** (embeddings), which enable **semantic search**—meaning the system can find similar documents based on meaning rather than just keywords.

Finally, a **FAISS vector store** is created from the processed documents using their embeddings. FAISS (Facebook AI Similarity Search) is a vector database optimized for fast and efficient similarity search, allowing quick retrieval of relevant information from large datasets.

Step 5: Define an Advanced Retrieval Function

In the following, we define an **advanced retrieval function** that finds the most relevant documents based on a user's query using **vector similarity search**.

The function `get_advanced_recommendations` takes three parameters:

1. **`query`**: The user's input or search phrase
2. **`k`**: The number of top matching results to return (default is 3)
3. **`return_scores`**: A boolean flag indicating whether to return similarity scores alongside the recommendations

The function calls `vector_store.similarity_search_with_score(query, k=k)`, which searches the FAISS vector store for the k most similar documents based on their embeddings. The results include both the retrieved document objects and their similarity scores.

If `return_scores` is `True`, the function returns a list of tuples containing both the document content and its similarity score. Otherwise, it returns only the document content without scores.

This approach enables **semantic search and recommendation generation**, making it useful for applications like **chatbots, document retrieval systems, and AI-powered recommendation engines** that need to find the most contextually relevant information.

```
# Define an advanced retrieval function
def get_advanced_recommendations(query, k=3, return_scores=False):
    """Returns the top-k most similar items based on user query, optionally
    with similarity scores."""
    results_with_scores = vector_store.similarity_search_with_
    score(query, k=k)
    if return_scores:
        return [(doc.page_content, score) for doc, score in results_
        with_scores]
    else:
        return [doc.page_content for doc, _ in results_with_scores]
```

Step 6: Integrate a QA System Using LangChain

Then, it's time to set up a **question-answering (QA) system** using LangChain by integrating a **retrieval-based approach** with an **LLM (GPT-4)**.

The `RetrievalQA.from_chain_type` function creates a **QA pipeline** that retrieves relevant information from a **vector store** before generating answers. It takes three key parameters:

1. **`llm=ChatOpenAI(model="gpt-4")`**: Uses OpenAI's **GPT-4** model to process and generate responses.
2. **`retriever=vector_store.as_retriever()`**: Converts the **FAISS vector store** into a retriever that can find the most relevant documents based on user queries.
3. **`chain_type="stuff"`**: Specifies the document processing method. The `"stuff"` method takes the retrieved documents, combines their content, and passes them directly to the LLM for response generation.

This setup enables **semantic search-based question answering**, where the system retrieves the most relevant documents from the vector store and leverages GPT-4 to **generate accurate, context-aware answers**. It's useful for **chatbots, document search engines, and AI assistants** that need to provide precise responses based on stored knowledge.

```
# Integrate a QA system using LangChain
qa_chain = RetrievalQA.from_chain_type(
    llm=ChatOpenAI(model="gpt-4"),
    retriever=vector_store.as_retriever(),
    chain_type="stuff"
)
```

Step 7: Set Up an AI Conversational Agent

The code below continues by setting up an **AI-powered conversational agent** using LangChain, integrating both a **question-answering (QA) system** and a **recommendation system** with conversational memory.

The `answer_query` function takes a user's input and retrieves an AI-generated answer using the `qa_chain.run(query)` method. This ensures responses are based on relevant retrieved information.

A **conversational memory** is implemented using `ConversationBufferMemory(memory_key="chat_history")`. This allows the chatbot to remember previous interactions, improving the coherence of multiturn conversations.

Two tools are defined using the `Tool` class:

1. **`recommendation_tool`**: Calls `get_advanced_recommendations()` to retrieve the **top five most relevant recommendations** based on a query, including similarity scores
2. **`qa_tool`**: Calls `answer_query()` to generate **AI-powered answers** based on retrieved documents

The **LangChain Agent** is then initialized using `initialize_agent()`:

- It includes the **QA and recommendation tools**.
- Uses **GPT-4** as the language model (`llm=ChatOpenAI(model="gpt-4")`).
- Implements a **zero-shot-react-description** agent, meaning the AI can reason and select the best tool dynamically.
- Maintains a conversation history using `memory`.
- Runs in **verbose mode**, providing detailed execution logs for debugging.

This setup enables the **AI agent to act as an intelligent assistant**, capable of both answering questions and providing recommendations in an interactive and memory-enhanced conversation.

```
# Function to answer user queries intelligently
def answer_query(query):
    """Returns an AI-generated answer based on retrieved information."""
    return qa_chain.run(query)

# Implement Conversational Memory
memory = ConversationBufferMemory(memory_key="chat_history")
```

```
# Define tools for the LangChain agent
recommendation_tool = Tool(
    name="Recommendation System",
    func=lambda query: get_advanced_recommendations(query, k=5, return_
    scores=True),
    description="Provides top recommendations based on a user query."
)

qa_tool = Tool(
    name="QA System",
    func=answer_query,
    description="Answers questions using an AI-powered retrieval system."
)

# Initialize LangChain Agent
agent = initialize_agent(
    tools=[recommendation_tool, qa_tool],
    llm=ChatOpenAI(model="gpt-4"),
    agent="zero-shot-react-description",
    memory=memory,
    verbose=True
)
```

Step 8: Test the System

```
# Example Queries
query = "I love sci-fi movies about space."
recommendations = get_advanced_recommendations(query, k=5, return_
scores=True)
print("Top Recommendations with Scores:")
for rec, score in recommendations:
    print(f"- {rec} (Score: {score:.4f})")

# Intelligent QA System Example
query_qa = "What are some movies about AI?"
answer = answer_query(query_qa)
print("\nAI-Powered Answer:")
print(answer)
```

```
# Interactive Agent Example
user_input = "Give me a recommendation for fantasy books."
agent_response = agent.run(user_input)
print("\nAgent Response:")
print(agent_response)
```

```
op Recommendations with Scores:
- Interstellar - A sci-fi movie about space exploration. (Score: 0.2047)
- The Matrix - A sci-fi movie about a simulated reality. (Score: 0.3131)
- Blade Runner - A dystopian sci-fi movie exploring artificial
  intelligence. (Score: 0.3341)
- Dune - A sci-fi novel by Frank Herbert about interstellar politics.
  (Score: 0.3552)
- Inception - A movie about dreams within dreams. (Score: 0.3911)
AI-Powered Answer:
Some movies about artificial intelligence include "Blade Runner" and "The
Matrix".
> Entering new AgentExecutor chain...
The user is asking for a recommendation, not a factual answer.
Action: Recommendation System
Action Input: Fantasy books
Observation: [('The Lord of the Rings - A fantasy novel by
J.R.R. Tolkien.', 0.2408208), ('The Witcher - A fantasy book and TV series
about a monster hunter.', 0.27408585), ("Harry Potter - A young wizard's
journey by J.K. Rowling.", 0.29327092), ('Dune - A sci-fi novel by Frank
Herbert about interstellar politics.', 0.35180575), ('Game of Thrones - A
TV series based on A Song of Ice and Fire.', 0.37026554)]
Thought:The recommendation system has provided a list of fantasy books.
There's no need to go further.
Final Answer: Here are some fantasy books you might enjoy: 'The Lord of
the Rings' by J.R.R. Tolkien, 'The Witcher' series, 'Harry Potter' series
by J.K. Rowling, 'Dune' by Frank Herbert, and the 'Game of Thrones' series
based on 'A Song of Ice and Fire'.
> Finished chain.
```

```
Agent Response:
Here are some fantasy books you might enjoy: 'The Lord of the Rings'
by J.R.R. Tolkien, 'The Witcher' series, 'Harry Potter' series by
J.K. Rowling, 'Dune' by Frank Herbert, and the 'Game of Thrones' series
based on 'A Song of Ice and Fire'.
```

App 12: PDF Files Chatbot

A **PDF Chatbot with LangChain** is an AI-powered assistant designed to interact with and extract insights from PDF documents. Using **LangChain**, a framework for building applications with large language models (LLMs), the chatbot can read, process, and answer questions based on the content of uploaded PDFs. This enables users to efficiently search for specific information, summarize sections, or analyze documents without manually going through large amounts of text.

Typically, a PDF chatbot integrates **text extraction tools** (like PyMuPDF or PDFMiner), **vector databases** for semantic search, and **LLM-powered reasoning** to provide accurate responses. This makes it useful for legal documents, research papers, contracts, and reports, improving workflow automation and knowledge retrieval.

Step 1: Install All Required Libraries

```
!pip install langchain pypdf faiss-cpu openai tiktoken
!pip install -U langchain-community
```

- **`langchain`**: The core framework that enables interaction with large language models (LLMs), document loading, vector databases, and reasoning capabilities.
- **`pypdf`**: A Python library for extracting text from PDF files, allowing the chatbot to read and process document content.
- **`faiss-cpu`**: A vector database library developed by Facebook AI for fast and efficient similarity search, crucial for storing and retrieving document embeddings.
- **`openai`**: Provides access to OpenAI's LLMs (like GPT-4) for natural language understanding and generating responses.

- **tiktoken**: A tokenizer used for counting tokens efficiently when working with OpenAI models, helping in cost estimation and ensuring token limits are managed properly.
- **langchain-community**: An updated version of LangChain's community-supported integrations. It includes various third-party tool integrations (like OpenAI, FAISS, Pinecone, and more) for better support and maintenance.

Then, import them:

```
import os
import faiss
import pickle
import time
from langchain.document_loaders import PyPDFLoader
from langchain.text_splitter import RecursiveCharacterTextSplitter
from langchain.embeddings.openai import OpenAIEmbeddings
from langchain.vectorstores import FAISS
from langchain.chains import ConversationalRetrievalChain
from langchain.chat_models import ChatOpenAI
from langchain.memory import ConversationBufferMemory
```

Step 2: Generate and Add Your OpenAI API Key

```
# Set API keys (Use environment variables for security)
import os
os.environ["OPENAI_API_KEY"] = <Your API Key>
```

Step 3: Upload Your PDF Files, Access Them, and Create a Vector Store Database

The following code sets up a system to process PDFs, extract text, split it into chunks, and store the processed data in a FAISS vector database for efficient retrieval. It first defines a directory named "**vector_store**" where the FAISS database will be stored.

The "**load_pdfs**" function takes a list of PDF file paths, extracts text from each file using "**PyPDFLoader**", and compiles all extracted text into a list. It then uses "**RecursiveCharacterTextSplitter**" to break the text into smaller chunks of 500 characters, ensuring a 100-character overlap between chunks for context preservation.

The "**create_or_load_vector_store**" function checks if a FAISS vector store already exists in the specified directory. If it does, it loads the stored embeddings using "**FAISS.load_local**". If not, it processes the PDFs by calling "**load_pdfs**", generates text embeddings using "**OpenAIEmbeddings**", and creates a new FAISS vector store. This new vector store is then saved locally for future use. The function returns the vector store, enabling efficient document search and retrieval using vector similarity.

```
# Directory to store vector database
DB_FAISS_PATH = "vector_store"

# Load and Process PDFs
def load_pdfs(pdf_paths):
    all_documents = []
    for pdf in pdf_paths:
        loader = PyPDFLoader(pdf)
        documents = loader.load()
        all_documents.extend(documents)

    # Split text into chunks
    text_splitter = RecursiveCharacterTextSplitter(chunk_size=500, chunk_
    overlap=100)
    chunks = text_splitter.split_documents(all_documents)
    return chunks

# Create or Load FAISS Vector Store
def create_or_load_vector_store(pdf_paths):
    if os.path.exists(DB_FAISS_PATH):
        print("[INFO] Loading existing vector store...")
        vectorstore = FAISS.load_local(DB_FAISS_PATH, OpenAIEmbeddings())
```

```
    else:
        print("[INFO] Creating new vector store...")
        chunks = load_pdfs(pdf_paths)
        vectorstore = FAISS.from_documents(chunks, OpenAIEmbeddings())
        vectorstore.save_local(DB_FAISS_PATH)
    return vectorstore
```

Step 4: Create a Chatbot with Memory

Then, we initialize a chatbot that can retrieve information from a vector store and engage in a conversation while maintaining memory. The **get_chatbot function** sets up a chatbot using the **GPT-4 model through ChatOpenAI.** It also initializes a **ConversationBufferMemory** to store the chat history, allowing the bot to remember previous exchanges within a session. The **ConversationalRetrievalChain** is then created, which enables the chatbot to retrieve relevant information from the vector store while keeping track of the conversation context.

The **chat_with_bot function** provides an interactive chat interface. It prints a message indicating that the chatbot is ready and waits for user input in a loop. If the user types "exit" or "quit," the loop breaks, and the program terminates the chat session. Otherwise, the user's query is passed to the **qa_chain**, which retrieves relevant information from the vector store and generates a response. The chatbot's reply is then printed, allowing for a continuous back-and-forth interaction.

```
# Initialize Chatbot with Memory
def get_chatbot(vectorstore):
    llm = ChatOpenAI(model_name="gpt-4")
    memory = ConversationBufferMemory(memory_key="chat_history", return_
    messages=True)
    qa_chain = ConversationalRetrievalChain.from_llm(
        llm, retriever=vectorstore.as_retriever(), memory=memory
    )
    return qa_chain

# Chat with the bot
def chat_with_bot(qa_chain):
    print("\n[INFO] Chatbot is ready! Type 'exit' to quit.")
```

```
while True:
    query = input("\nYou: ")
    if query.lower() in ["exit", "quit"]:
        print("\n[INFO] Exiting chat...\n")
        break
    response = qa_chain.run(query)
    print(f"Bot: {response}")
```

Step 5: Ask the Chatbot and Receive an Answer

```
# Main function
if __name__ == "__main__":
    pdf_files = ["Jira_Software.pdf", "ML+Cheat+Sheet_2.pdf"]  # Replace
    with your PDFs
    vectorstore = create_or_load_vector_store(pdf_files)
    chatbot = get_chatbot(vectorstore)
    chat_with_bot(chatbot)
```

Output:

```
[INFO] Creating new vector store...
<ipython-input-6-5c12dafc1a0e>:25: LangChainDeprecationWarning: The class
`OpenAIEmbeddings` was deprecated in LangChain 0.0.9 and will be removed
in 1.0. An updated version of the class exists in the :class:`~langchain-
openai package and should be used instead. To use it run `pip install -U
:class:`~langchain-openai` and import as `from :class:`~langchain_openai
import OpenAIEmbeddings``.
  vectorstore = FAISS.from_documents(chunks, OpenAIEmbeddings())
<ipython-input-7-ec9b7022866c>:3: LangChainDeprecationWarning: The class
`ChatOpenAI` was deprecated in LangChain 0.0.10 and will be removed in
1.0. An updated version of the class exists in the :class:`~langchain-
openai package and should be used instead. To use it run `pip install -U
:class:`~langchain-openai` and import as `from :class:`~langchain_openai
import ChatOpenAI``.
```

```
  llm = ChatOpenAI(model_name="gpt-4")
<ipython-input-7-ec9b7022866c>:4: LangChainDeprecationWarning: Please
see the migration guide at: https://python.langchain.com/docs/versions/
migrating_memory/
  memory = ConversationBufferMemory(memory_key="chat_history", return_
  messages=True)
[INFO] Chatbot is ready! Type 'exit' to quit.

You: What is the model in the Jira software pdf?
<ipython-input-7-ec9b7022866c>:18: LangChainDeprecationWarning: The method
`Chain.run` was deprecated in langchain 0.1.0 and will be removed in 1.0.
Use :meth:`~invoke` instead.
  response = qa_chain.run(query)
Bot: The model in the JIRA software PDF is a deep learning system designed
to automate the categorization and prioritization of JIRA tickets. It aims
to accurately extract relevant insights from ambiguous text, classify
tickets based on urgency and relevance, and eliminate the manual triaging
effort. The input to the model consists of a sequence of discrete tokens,
represented as an integer vector. The study compares different approaches
to address the complexity of JIRA ticket classification.

You: exit

[INFO] Exiting chat...
```

Summary

In Chapter 3, we explored the development of advanced applications powered by large language models (LLMs) using LangChain and Python. We examined how LangChain provides a modular approach to building AI-driven solutions, enabling capabilities like multistep reasoning, dynamic interactions, and seamless data integration. Through practical implementations, such as YouTube video summarization and intelligent document analysis, we demonstrated how LLMs can be applied to real-world problems. Additionally, we addressed key challenges, including optimizing model behavior, handling errors, and ensuring application scalability in production environments.

With a strong foundation in application development, we now shift our focus to deployment strategies in Chapter 4. Building LLM-powered applications is just the beginning—successfully deploying them at scale requires careful consideration of infrastructure, optimization, and performance management. This chapter explores cloud deployment strategies, memory and computational efficiency techniques, and best practices for ensuring security and scalability. By understanding the complexities of deploying LLM applications, you will be equipped to transition from development to real-world implementation, making AI-powered solutions accessible and efficient in production environments.

CHAPTER 4

Deploying LLM-Powered Applications

The deployment of large language models (LLMs) marks a pivotal step in transforming cutting-edge AI research into impactful real-world applications. Whether enabling conversational agents, automating content creation, or driving decision-making tools, LLMs unlock opportunities for innovation across industries. However, deploying these powerful models is far from straightforward. It requires navigating a landscape of technical challenges, architectural choices, and optimization techniques to ensure performance, scalability, and efficiency in production environments.

This chapter focuses on the critical aspects of deploying LLM-powered applications, equipping you with the knowledge to tackle this complex process effectively. We will begin with an exploration of cloud deployment strategies and scalability considerations, essential for ensuring that your application can handle varying loads and user demands. Building on this foundation, we'll delve into best practices for deploying LLMs in production, highlighting strategies that balance speed, cost, and reliability.

Next, we'll explore the tools available for deploying LLMs, from infrastructure frameworks to model-serving solutions, providing a comprehensive toolkit to simplify and streamline the process. As with any cutting-edge technology, deploying LLMs comes with its challenges. From inference latency and memory constraints to managing large-scale infrastructure, we will identify the hurdles you may encounter and propose solutions to address them.

Optimization plays a central role in deploying LLMs effectively. We'll examine memory optimization techniques and compression strategies that can reduce resource usage without compromising model performance. Additionally, we'll investigate advanced techniques for optimizing attention layers, a critical component of LLMs, and explore scheduling optimizations at various levels—request, batch, and iteration—to enhance throughput and responsiveness.

D. Grigorov, *Intermediate Python and Large Language Models*, https://doi.org/10.1007/979-8-8688-1475-4_4

This chapter outlines the core strategies, tools, and challenges involved in deploying large language models at scale:

- **Overview**: Why LLM deployment matters and key challenges
- **Deployment**: Cloud setup, scalability, performance, and cost
- **Integration**: Hosted, prepackaged, and open source models
- **Tools**: Frameworks, platforms, and example workflows
- **Optimization**: Compression, attention, memory, and scheduling
- **Challenges**: Latency, scaling, reliability, and efficiency
- **Takeaways**: Best practices for scalable, ethical deployment

By the end of this chapter, you'll gain a clear understanding of the strategies and tools necessary to deploy LLMs at scale, overcoming technical barriers while ensuring your applications meet the demands of users and stakeholders. Deploying LLM-powered applications is a multifaceted challenge, but with the right knowledge and approach, you can turn these models into practical, high-performing solutions that deliver real value. Let's dive in.

Note While this chapter includes examples and tools, its purpose is not to provide a definitive framework for deploying your LLM application. Each use case is unique and requires a tailored approach.

Integrating LLMs into Web and Mobile Applications

Large language models (LLMs) have revolutionized how applications handle language understanding and generation, opening up possibilities for automating complex tasks, improving efficiency, and enhancing user experiences. From content creation and sentiment analysis to answering queries and driving conversational AI, LLMs can transform a variety of industries. However, integrating LLMs into your workflow requires thoughtful planning, as the method you choose will impact costs, scalability, customization, and privacy. Below, we explore three primary approaches to integrating LLMs: hosted models, prepackaged solutions, and deploying open source models.

Hosted Models

Hosted models offer the quickest and easiest way to access the capabilities of LLMs. Companies like OpenAI, Google Cloud Platform (GCP), and Azure provide hosted services, allowing users to interact with pretrained models like GPT or Gemini through APIs. This option eliminates the need for infrastructure setup or maintenance and allows even nontechnical teams to implement advanced AI features.

How Hosted Models Work

Hosted models operate via API interfaces. Developers send requests (or "prompts") to the service provider's server and receive the model's response. These APIs are typically well-documented and designed to be user-friendly, enabling seamless integration with existing systems.

Advantages of Hosted Models

- **No Setup Required**: Hosted models require no installation, infrastructure configuration, or optimization. This makes them ideal for teams with limited technical resources or those needing quick deployment.
- **Scalability Managed by Providers**: Cloud providers automatically scale resources to meet usage demands, ensuring smooth operation during peak loads.
- **Simplified Interfaces**: APIs abstract away technical complexities, making it easy to send text prompts and receive model responses.
- **Rich Tooling Ecosystem**: Hosted model providers offer a wide range of tools that streamline development, orchestration, and integration. For example:
    - **Anthropic's Model Context Protocol (MCP)**: Enables advanced context management when using Claude in multiagent or tool-augmented setups.
    - **OpenAI Function Calling and Assistant API**: Allows developers to define tools/functions the model can invoke, making it easier to build agents and tool-using workflows.

- **LangChain and LlamaIndex Integrations**: Many providers support or offer integrations with popular frameworks for chaining model calls, retrieval-augmented generation (RAG), and memory handling.
- **Azure OpenAI Studio and Playground**: Provides a GUI for model testing, prompt engineering, and deployment configuration directly from the cloud console.
- **Google Vertex AI Extensions for Gemini**: Supports building multimodal workflows, tool integration, and connecting to enterprise data sources.

Challenges of Hosted Models

- **Cost**: Usage fees are based on API calls or data processed, and costs can escalate with high-volume applications.
- **Limited Customization**: Hosted models are "as-is," meaning users cannot fine-tune them for niche applications or domain-specific needs.
- **Data Privacy Concerns**: Sending sensitive or proprietary data to a third-party server—such as through a hosted model or an external API like OpenAI's—can be risky, particularly in sensitive industries like finance, healthcare, or legal services. However, if you self-host an LLM on a VPN-enabled server, many of these data privacy concerns can be significantly mitigated, since the data remains within your controlled environment.

Hosted models are best suited for projects with minimal customization needs, moderate budgets, and tight timelines. They are also a great choice for prototyping and proof-of-concept work, allowing teams to experiment with LLM capabilities before committing to more complex implementations.

Prepackaged Models

Prepackaged models provide a balance between ease of use and control. These are pretrained language models offered by platforms like Hugging Face, optimized and bundled with essential tools for deployment. Designed to simplify the deployment process, prepackaged models allow users to leverage advanced AI while retaining more control over their infrastructure.

Components of Prepackaged Models

- **Model Selection**: Models are pretrained on large datasets and fine-tuned for specific domains, enabling applications in areas like customer support, healthcare, or finance.
- **Optimization**: To enhance performance and efficiency, prepackaged models are optimized using techniques such as
    - **Quantization**: Reduces memory usage and speeds up inference by converting model parameters to lower precision formats
    - **Pruning**: Removes redundant parameters, reducing model size without significantly affecting accuracy
    - **Distillation**: Creates a smaller "student" model trained to mimic the larger model's behavior, improving efficiency for deployment
- **Bundled Software**: These models come with preintegrated software components such as
    - **Inference Engines**: Optimize the execution of model computations
    - **APIs or SDKs**: Provide user-friendly interfaces for developers to interact with the model
    - **Deployment Scripts**: Facilitate the installation and configuration of models on different platforms.
    - **Documentation**: Includes detailed guides on setup, usage, and troubleshooting

Advantages of Prepackaged Models

- **Better Control**: Compared to hosted models, prepackaged models allow users to fine-tune and optimize for specific use cases.
- **Data Privacy**: Models can be deployed on private infrastructure, ensuring sensitive data never leaves the organization's systems.
- **Streamlined Setup**: Prepackaged solutions simplify what could otherwise be a highly complex deployment process.

Challenges of Prepackaged Models

- **Technical Expertise Needed**: While more accessible than open source models, prepackaged solutions still require some familiarity with infrastructure setup and maintenance.
- **Upfront Investment**: Infrastructure and initial deployment may require financial and resource investment.

Prepackaged models are ideal for organizations looking to maintain some control over data and customization while leveraging ready-made tools to simplify deployment.

Deploying Open Source Models

Open source models provide the highest level of control and flexibility. Developers download model weights (parameters) and adapt the models to their unique requirements. Open source solutions are often shared through repositories like Hugging Face, providing a wide range of options from lightweight models to highly advanced LLMs.

Steps to Deploy Open Source Models

- **Model Selection**: Choose a model that aligns with your application's goals, such as accuracy, efficiency, or resource constraints.
- **Download the Model**: Use repositories like Hugging Face's Transformers library to access and load the model.

- **Environment Setup**: Configure hardware (GPUs, TPUs) and software environments. Popular frameworks include
    - **Text Generation Inference (TGI)**: Optimized for large-scale text generation
    - **Transformer Agents**: For specific applications requiring complex workflows

Model Deployment

- **Local Deployment**: For testing or small-scale applications, a local environment is sufficient.
- **Cloud Deployment**: For large-scale use, containerization (e.g., Docker) is often employed to manage dependencies and streamline deployment.

Advantages of Open Source Deployment

- **Maximum Control**: Users can customize, fine-tune, and modify models to suit specific needs.
- **Data Privacy**: By deploying models on local or private infrastructure, organizations maintain complete control over sensitive data.
- **Cost Efficiency**: While initial setup costs may be high, eliminating API usage fees can lead to significant long-term savings.

Challenges of Open Source Deployment

- **High Technical Expertise Required**: Teams must have a strong background in machine learning, model optimization, and infrastructure management.
- **Complex Setup**: Deployment requires significant time and effort, especially for large models that demand high computational resources.
- **Maintenance**: Ongoing updates and optimizations are necessary to ensure the model remains performant and efficient.

Open source deployment is best suited for organizations with advanced technical capabilities and a need for tailored solutions.

Factors to Consider When Choosing a Method

When deciding how to integrate LLMs into your applications, consider these key factors:

Technical Expertise

- Hosted models are ideal for beginners or teams with limited technical skills.
- Prepackaged models require moderate technical expertise.
- Open source models demand advanced skills in machine learning and infrastructure management.

Data Privacy

- If handling sensitive or proprietary data, avoid hosted models where data is transmitted to third-party servers.
- Prepackaged and open source models deployed on private infrastructure offer greater privacy.

Cost

- Hosted models involve ongoing operational costs tied to usage.
- Prepackaged models balance initial setup costs with manageable long-term expenses.
- Open source models require significant upfront investment in infrastructure but eliminate recurring API fees.

Scalability

- Hosted models handle scaling automatically.
- Prepackaged solutions often include tools for scaling in cloud environments.
- Open source models require custom scaling solutions, increasing complexity.

Customization Needs

- Hosted models provide limited customization.
- Prepackaged and open source models enable significant customization for domain-specific tasks.

Integrating large language models into your applications can unlock powerful capabilities, but the method you choose depends on your specific needs, resources, and constraints. Hosted models offer unparalleled simplicity and scalability but come with ongoing costs and privacy trade-offs. Prepackaged models provide a middle ground, offering ease of deployment with more control over customization. Open source models give full flexibility and privacy but require significant technical expertise and effort.

By carefully evaluating your project's goals, budget, and technical capacity, you can select the method that aligns best with your objectives. Whether you prioritize speed to market, control over data, or long-term cost savings, there is an approach to fit your needs, enabling you to leverage the transformative power of LLMs.

LLM Cloud Deployment and Scalability Considerations

Deploying and scaling a large language model (LLM) in the cloud is a multifaceted process that requires thorough planning, precise execution, and ongoing management. To ensure efficient performance, high availability, and cost-effectiveness, several aspects must be carefully considered. Below is an in-depth exploration of these considerations.

Deployment Architecture

The architecture of an LLM deployment forms the foundation of its performance and scalability. Key architectural considerations include the use of load balancers to distribute incoming requests evenly across model instances. This ensures that no single instance becomes a bottleneck and enhances system reliability. Cloud-native load balancers, such as those provided by AWS, Google Cloud, and Microsoft Azure, are well-suited for this purpose.

Auto-scaling is another essential feature, enabling the infrastructure to dynamically adjust the number of model instances based on request volume and latency metrics. This ensures optimal resource utilization and cost-effectiveness during periods of fluctuating demand. Advanced auto-scaling setups might involve predictive scaling, where machine learning models forecast demand based on historical data, allowing for preemptive scaling to avoid latency spikes.

Caching frequently requested responses can significantly reduce computational load and improve response times. For example, implementing a cache layer for common queries ensures that these can be served without invoking the full inference pipeline. Leveraging distributed cache systems such as Redis or Memcached can add scalability and reliability to the caching layer.

Queue systems for asynchronous processing are valuable for handling workloads where immediate responses are not required. These systems decouple request submission from processing, allowing for better resource management during peak traffic periods. For example, tasks like batch translations or large document summarizations can be offloaded to a message queue system like RabbitMQ or AWS SQS, ensuring seamless operation even during high-demand periods.

Infrastructure

The infrastructure supporting LLM deployments must be optimized for high-performance inference. **GPU clusters** are essential for handling the computational demands of LLMs, particularly during inference. **Monitoring GPU utilization** ensures that resources are effectively used and identifies underutilized instances for cost savings. Advanced GPU resource management might involve GPU pooling or dynamic resource reallocation to ensure maximum efficiency.

Memory and storage optimization is critical for managing large model weights. Techniques such as model sharding, where weights are distributed across multiple devices, can help accommodate larger models. Additionally, ensuring sufficient storage bandwidth and capacity minimizes bottlenecks during inference. Employing high-speed NVMe storage or direct-attached storage (DAS) can provide the necessary throughput for data-intensive operations.

Network capacity planning is another vital consideration. High-throughput inference requires robust networking to minimize latency and ensure smooth data flow between components. Using software-defined networking (SDN) or high-bandwidth interconnects can further enhance network performance. Employing container orchestration tools like Kubernetes streamlines the deployment process, providing scalability, fault tolerance, and simplified management. Kubernetes operators designed for AI workloads, such as Kubeflow, can further enhance the efficiency of managing LLM deployments.

Performance

Optimizing the performance of LLM systems involves several strategies. Model quantization reduces the precision of weights and activations (e.g., from FP32 to INT8), lowering computational requirements and speeding up inference without significantly affecting accuracy. Similarly, model distillation creates smaller, efficient models that replicate the performance of larger ones. These methods not only improve performance but also reduce infrastructure costs.

Batching requests is an effective way to maximize GPU utilization. By processing multiple requests simultaneously, batching reduces overhead and increases throughput. This approach is especially effective in high-demand environments, such as customer support systems or real-time recommendation engines.

Response streaming allows the system to deliver initial tokens of a response while generating subsequent tokens. This approach improves perceived latency and is particularly useful for conversational applications, such as chatbots or virtual assistants. Integrating response streaming with adaptive pacing algorithms can further refine user experience by dynamically adjusting token delivery based on network conditions and user interaction.

Load testing is critical for identifying performance bottlenecks and ensuring that the system can handle expected traffic volumes. Tools like Locust or JMeter can simulate workloads and provide actionable insights. More advanced testing setups might involve chaos engineering techniques, where intentional disruptions are introduced to test the system's resilience under failure scenarios.

Cost Management

Effective cost management ensures the sustainability of LLM deployments. Instance right-sizing involves selecting hardware configurations that align with workload patterns. Overprovisioning resources can lead to unnecessary expenses, while underprovisioning can impact performance. Regular audits of resource utilization can identify opportunities to optimize costs.

Spot instances, which offer spare cloud capacity at reduced prices, are ideal for noncritical workloads or batch processing. However, these instances can be preempted, so they should be used with failover mechanisms. Employing checkpointing techniques allows for intermediate progress to be saved, minimizing the impact of instance termination.

Multiregion deployments reduce latency by bringing resources closer to end users. This approach also enhances availability by providing redundancy in case of regional outages. Utilizing cost-efficient regions for noncritical workloads can further optimize expenses without compromising service quality.

Resource allocation based on priority tiers ensures that critical workloads receive the necessary resources, while lower-priority tasks are executed with cost-saving measures. Implementing tiered resource allocation policies can streamline budgeting and operational efficiency.

Monitoring

Robust monitoring practices are essential for maintaining the health and performance of LLM systems. Key metrics to monitor include inference latency, throughput, error rates, and resource utilization across the stack. Tools like Prometheus, Grafana, and cloud-native monitoring services can provide real-time visibility into these metrics. For more granular monitoring, integrating AI-focused observability tools like MLFlow or SageMaker Monitor can track model-specific performance indicators.

Monitoring model performance is crucial for detecting degradation over time. Regular evaluations can identify when retraining or fine-tuning is needed. Drift detection mechanisms can flag changes in input data distribution that may affect model accuracy, prompting timely intervention.

Cost per inference tracking provides insights into the economic efficiency of the deployment, helping teams identify opportunities for optimization. Establishing alerts for anomalies in resource utilization or costs ensures proactive issue resolution. Additionally, employing predictive analytics can forecast resource requirements, aiding in more strategic planning.

High Availability and Fault Tolerance

To ensure reliability, LLM deployments must be designed for high availability and fault tolerance. Deploying resources across multiple regions provides redundancy and ensures that services remain available even during regional outages. Advanced configurations might involve active-active setups, where multiple regions actively serve requests, further enhancing reliability and reducing latency.

Regular backups of model weights, configurations, and data are essential for disaster recovery. Automated recovery mechanisms should be in place to restore services quickly in case of failures. Implementing retry logic in the communication between components can address transient errors and enhance overall reliability. For critical workloads, employing consensus protocols like Raft or Paxos can ensure consistent state management across distributed systems.

Compliance and Ethics

Compliance with data privacy regulations, such as GDPR or CCPA, is essential when deploying LLMs. This involves securing user data, obtaining necessary consents, and implementing robust data governance policies. Leveraging privacy-preserving techniques such as differential privacy or federated learning can further enhance compliance.

Bias mitigation is another critical consideration. Regular audits of the model's behavior can help detect and reduce biases, ensuring fair and ethical outcomes. Incorporating fairness metrics into the development pipeline can provide ongoing insights into model behavior. Transparency about the model's limitations and behavior builds trust with users and stakeholders. Creating detailed documentation and user guides about model use cases and potential risks enhances accountability.

Deploying and scaling LLMs in the cloud is a complex but rewarding endeavor. By carefully considering deployment architecture, infrastructure, performance, cost management, monitoring, high availability, and compliance, organizations can create reliable, efficient, and ethical solutions. Continuous improvement and adaptation to emerging technologies and challenges will ensure long-term success in leveraging the power of LLMs. In this rapidly evolving field, staying informed and proactive will be key to maintaining competitive advantage and delivering value to users.

Tools for Deploying LLMs

Model Hosting Frameworks

Frameworks and libraries provide the foundation for hosting and serving machine learning models, enabling developers to create robust APIs and interfaces.

- **Hugging Face Transformers**: One of the most popular libraries for working with LLMs. It supports pretrained models for tasks like text generation, summarization, and more. The library includes extensive integration with other tools for fine-tuning and deployment.
- **Hugging Face Accelerate**: Simplifies the deployment of models on distributed systems and multi-GPU setups. It integrates seamlessly with Hugging Face Transformers, making it ideal for scaling up training or inference.
- **FastAPI**: A modern, high-performance web framework for Python that allows developers to quickly create APIs for exposing LLM functionalities. Its asynchronous capabilities make it highly suitable for LLM inference.
- **Flask**: Lightweight and simple, Flask is often used for prototyping and building small-scale APIs to serve models.
- **TorchServe**: Specifically designed for PyTorch models, TorchServe offers features like batch inference, metrics tracking, and customizable handlers for complex preprocessing or postprocessing tasks.
- **TensorFlow Serving**: A powerful system for serving TensorFlow models at scale, with built-in support for versioning and A/B testing of deployed models.
- **Gradio**: A low-code framework to create user interfaces for LLMs. It's perfect for building demos or interactive applications for text generation, question answering, or other LLM tasks.
- **Streamlit**: Similar to Gradio, but more focused on building dashboards and interactive data-driven applications for LLM outputs.
- **BentoML**: Offers an end-to-end workflow for deploying and serving machine learning models. It supports multiple back ends and provides a unified interface for deployment.
- **TRITON Inference Server**: Developed by NVIDIA, Triton supports multiple machine learning frameworks (e.g., PyTorch, TensorFlow, ONNX) and offers GPU-optimized inference pipelines.

Example: Saving a Model Locally, Uploading It to Hugging Face, and Calling It

1. **Install transformers and huggingface_hub**: type: `pip install transformers==4.50.3 huggingface_hub==0.30.1`
2. **Log in to Hugging Face**: type: `huggingface-cli login`
3. **Save your model locally.** For this example, let's save a pretrained distilbert-base-uncased model.

```
from transformers import AutoModelForSequenceClassification,
AutoTokenizer
# Load the model and tokenizer
model_name = "distilbert-base-uncased"
model = AutoModelForSequenceClassification.from_
pretrained(model_name)
tokenizer = AutoTokenizer.from_pretrained(model_name)

# Save the model and tokenizer locally
model.save_pretrained("./my_model")
tokenizer.save_pretrained("./my_model")
```

4. **Upload the model to Hugging Face**

```
from huggingface_hub import upload_folder

from huggingface_hub import create_repo
create_repo(repo_name)
# Define your repository name
repo_name = "your-username/my-first-model"

# Upload the model directory
upload_folder(
    folder_path="./my_model",
    repo_id=repo_name,
    commit_message="Initial model upload"
)
```

```
print(f"Model uploaded to https://huggingface.co/{repo_name}")
```

Your model will now be available at https://huggingface.co/your-username/my-first-model.

5. **Calling the model**

```
from transformers import AutoModelForSequenceClassification,
AutoTokenizer

# Load the model from Hugging Face
model_name = "your-username/my-first-model"
tokenizer = AutoTokenizer.from_pretrained(model_name)
model = AutoModelForSequenceClassification.from_
pretrained(model_name)

# Example text
input_text = "Hugging Face makes working with AI easy and fun!"
inputs = tokenizer(input_text, return_tensors="pt")

# Get predictions
outputs = model(**inputs)
print(outputs.logits)  # Logits for classification
```

Output:

It depends on the specific task your model is designed to solve.

Optimization Tools

Optimization tools are essential for reducing inference latency and memory usage, especially when deploying large models.

- **ONNX (Open Neural Network Exchange)**: Converts models to an open format that can run on optimized runtimes across various hardware architectures. ONNX is a critical step for deploying LLMs on diverse platforms.
- **ONNX Runtime**: An execution engine for ONNX models that accelerates inference through hardware-specific optimizations.

- **DeepSpeed**: Designed for both training and inference, DeepSpeed offers features like ZeRO optimization to handle memory-intensive LLMs, making it possible to train and deploy models with limited resources.
- **NVIDIA TensorRT**: Provides GPU-accelerated inference by optimizing neural networks, particularly effective for transformer-based architectures.
- **Hugging Face Optimum**: A library that bridges Hugging Face Transformers with optimized inference techniques using ONNX, TensorRT, and other acceleration technologies.
- **Intel Neural Compressor**: Specializes in quantizing models to lower precision (e.g., INT8) for faster inference on Intel processors.
- **BitsAndBytes**: A tool for quantizing large models down to as low as 4-bit precision, ideal for reducing resource demands without significant performance loss.
- **OpenVINO**: An Intel toolkit that optimizes and deploys models for CPUs, GPUs, and edge devices, suitable for use cases where LLMs need to run in constrained environments.
- **TVM/Apache TVM**: A deep learning compiler stack that automates optimization and deployment across a wide range of hardware platforms.
- **vLLM** is a high-performance inference and serving engine designed to optimize the deployment of large language models (LLMs). It addresses common performance bottlenecks such as inefficient memory usage, high latency, and limited throughput under concurrent workloads.

  vLLM stands for virtualized LLM. It is an open source project developed to support fast, efficient, and scalable LLM inference, particularly in production environments or applications with high traffic and real-time demands.

- **Key Optimizations and Features**
    - **PagedAttention Mechanism**: Traditional inference systems allocate fixed memory blocks for each request, often leading to fragmentation and underutilization. vLLM introduces PagedAttention, a dynamic memory management scheme that allocates attention key/value caches more efficiently. This allows for better memory utilization, the ability to serve many concurrent requests, and improved performance in real-time scenarios.
    - **High Throughput and Low Latency**: vLLM is designed to minimize token-level processing overhead, enabling fast generation and response times even with large models. This makes it particularly well-suited for applications that rely on streaming outputs, such as chat interfaces or interactive assistants.
    - **Multitenancy and Session Management**: vLLM supports multiple simultaneous sessions by virtualizing GPU memory usage. This enables multiple users or model endpoints to share GPU resources without the need for duplicating model weights or running separate processes.
    - **OpenAI-Compatible API**: vLLM provides an OpenAI-compatible API interface, making it easy to integrate with existing services and tools that rely on OpenAI's format. This allows for quick migration or testing without significant changes to the frontend or client infrastructure.
    - **Integration and Ecosystem**: vLLM is compatible with Hugging Face Transformers and can leverage additional performance enhancements through integrations with FlashAttention, DeepSpeed, and Triton kernels. This flexibility makes it a strong choice for both research and production use.

ONNX Example

1. Install required libraries

```
pip install torch transformers onnx onnxruntime
```

2. Export a PyTorch model to ONNX. We will use a Hugging Face transformer model (e.g., distilbert-base-uncased) and save it in ONNX format.

```
import torch
from transformers import AutoTokenizer,
AutoModelForSequenceClassification

# Load model and tokenizer
model_name = "distilbert-base-uncased"
tokenizer = AutoTokenizer.from_pretrained(model_name)
model = AutoModelForSequenceClassification.from_
pretrained(model_name)

# Example input
text = "Hugging Face makes AI accessible."
inputs = tokenizer(text, return_tensors="pt")

# Export the model to ONNX
torch.onnx.export(
    model,                                              # Model
    (inputs["input_ids"], inputs["attention_mask"]),  # Input
                                                        arguments
    "model.onnx",                                      # Output file
    input_names=["input_ids", "attention_mask"], # Input names
    output_names=["logits"],                           # Output name
    dynamic_axes={"input_ids": {0: "batch_size"}, "attention_
    mask": {0: "batch_size"}},  # Dynamic batch
    opset_version=11                                   # ONNX
    opset version
)

print("Model exported to ONNX format as 'model.onnx'")
```

3. **Run the ONNX model using ONNX Runtime. Load the ONNX model, and perform inference using the ONNX Runtime library**:

```
import onnxruntime as ort
import numpy as np

# Load the ONNX model
onnx_model_path = "model.onnx"
ort_session = ort.InferenceSession(onnx_model_path)

# Tokenize the input text
inputs = tokenizer(text, return_tensors="np")  # Use NumPy format
                                               for ONNX Runtime

# Prepare inputs
input_ids = inputs["input_ids"]
attention_mask = inputs["attention_mask"]

# Run inference
outputs = ort_session.run(
    None,  # Output names (None means all outputs)
    {"input_ids": input_ids, "attention_mask": attention_mask},  #
    Input dictionary
)

# Extract logits
logits = outputs[0]
print("Logits:", logits)
```

Output:

It depends on the specific task your model is designed to solve.

Cloud Services

Cloud platforms provide the necessary compute resources and infrastructure to host LLMs at scale.

- **AWS SageMaker**: An end-to-end machine learning platform with tools for training, tuning, and deploying LLMs. SageMaker endpoints allow for seamless integration with other AWS services.
- **Google Cloud Vertex AI**: A managed service that supports training and deploying large models with TPU integration for high performance.
- **Microsoft Azure**: Offers the OpenAI Service, allowing users to leverage GPT models like GPT-4 and Codex directly within their applications.
- **IBM Watson Studio**: Focuses on enterprise-grade AI, providing tools for building, deploying, and managing large-scale AI applications.
- **Lambda Labs**: Specializes in high-performance GPUs for training and serving LLMs, ideal for teams requiring raw computational power.
- **Paperspace Gradient**: Simplifies LLM workflows with preconfigured infrastructure and tools for collaborative model development.
- **Replicate**: Provides hosted APIs for deploying pretrained models with minimal configuration.
- **Modal**: Allows seamless deployment of machine learning pipelines to the cloud with support for GPUs and scalable infrastructure.

AWS SageMaker Example

1. **Install the required libraries: pip install boto3 sagemaker transformers.**

   Ensure you have an AWS account and the AWS CLI configured with appropriate permissions to use SageMaker.

2. **Upload a pretrained model to S3.**

```
import boto3
from transformers import AutoModelForSequenceClassification,
AutoTokenizer
```

```
# AWS setup
bucket_name = "your-s3-bucket-name"  # Replace with your S3
                                       bucket name
prefix = "models/bert"              # Folder path in the bucket
s3 = boto3.client("s3")

# Load pre-trained model
model_name = "distilbert-base-uncased"
model = AutoModelForSequenceClassification.from_
pretrained(model_name)
tokenizer = AutoTokenizer.from_pretrained(model_name)

# Save model locally
model.save_pretrained("./model")
tokenizer.save_pretrained("./model")

# Upload to S3
s3.upload_file("./model/config.json", bucket_name, f"{prefix}/
config.json")
s3.upload_file("./model/pytorch_model.bin", bucket_name,
f"{prefix}/pytorch_model.bin")
s3.upload_file("./model/tokenizer_config.json", bucket_name,
f"{prefix}/tokenizer_config.json")
s3.upload_file("./model/vocab.txt", bucket_name, f"{prefix}/
vocab.txt")
print(f"Model uploaded to S3: s3://{bucket_name}/{prefix}")
```

3. **Deploy the model on SageMaker. Use SageMaker to deploy the model as an endpoint.**

```
import sagemaker
from sagemaker.huggingface import HuggingFaceModel

# Specify the model's S3 location
model_data = f"s3://{bucket_name}/{prefix}"

# Define the Hugging Face model parameters
huggingface_model = HuggingFaceModel(
    model_data=model_data,
```

```
    role="your-sagemaker-execution-role",  # Replace with your IAM
                                            role for SageMaker
    transformers_version="4.17",           # Adjust based on the
                                           model's version
    pytorch_version="1.10",
    py_version="py38",
)

# Deploy the model as a SageMaker endpoint
predictor = huggingface_model.deploy(
    initial_instance_count=1,
    instance_type="ml.m5.large",  # Instance type for hosting
)
print("Model deployed to SageMaker!")
```

4. **After deploying the model, you can send data to the endpoint for inference.**

```
# Example text
input_text = "SageMaker makes deploying ML models easy!"

# Prepare the input for the model
data = {"inputs": input_text}

# Send the data to the deployed endpoint
response = predictor.predict(data)

# Print the model's prediction
print("Model Prediction:", response)
```

Orchestration and Scaling

As deployments grow, orchestration and scaling tools help manage complexity and ensure reliability.

- **Kubernetes**: A container orchestration system for managing distributed applications. It's widely used for scaling LLM deployments in production.

- **Ray Serve**: A scalable model serving library built on the Ray framework, suitable for distributed inference workloads.
- **Kubeflow**: A Kubernetes-native platform for building and deploying end-to-end machine learning workflows.
- **MLflow**: A tool for tracking experiments, packaging models, and managing deployments. MLflow simplifies version control and collaborative workflows.
- **Airflow**: Workflow orchestration tool to automate the deployment and monitoring of LLM pipelines.
- **Argo Workflows**: Provides Kubernetes-native workflows for automating complex multistep processes.

Edge and Mobile Deployment

Deploying LLMs to edge devices ensures low-latency inference and privacy preservation.

- **TensorFlow Lite**: Optimizes TensorFlow models for mobile and embedded systems, with support for hardware acceleration
- **PyTorch Mobile**: Enables PyTorch models to run on mobile devices, supporting custom optimizations
- **NVIDIA Jetson Platform**: Combines hardware and software for deploying LLMs on edge devices with GPU acceleration
- **CoreML**: Apple's framework for running machine learning models on iOS/macOS devices
- **Edge Impulse**: Simplifies deploying LLMs to constrained edge hardware for industrial applications

APIs for Hosted Models

For developers who prefer using hosted solutions, APIs offer a quick way to access powerful LLMs.

- **OpenAI API**: Provides access to GPT models for a wide range of applications, from chatbots to text summarization

- **Cohere API**: Focused on NLP tasks like embeddings, classification, and generation
- **Anthropic Claude API**: Offers conversational models with an emphasis on safety and alignment
- **AI21 Labs API**: Provides robust LLMs like Jurassic for various text-processing tasks

Distributed Inference and Fine-Tuning

Handling large models across multiple nodes or GPUs requires specialized tools.

- **DeepSpeed-Inference**: Optimized for scaling LLM inference across distributed systems
- **Alpa**: Automates parallelization strategies for large-scale models
- **FlexGen**: Enables efficient inference of large models on limited hardware
- **FasterTransformer**: NVIDIA's library for high-speed transformer model inference

Monitoring and Observability

To ensure models perform well in production, monitoring tools are essential.

- **Prometheus**: Collects real-time metrics for system and model monitoring
- **Grafana**: Visualizes performance metrics in interactive dashboards
- **Datadog**: Offers observability tools for tracking model and system health
- **Weights & Biases (W&B)**: Tracks experiments and monitors deployed models

LLM Inference Challenges: A Comprehensive Exploration

Deploying large language models (LLMs) for inference has become one of the most pressing challenges in modern AI systems. As these models grow in size and complexity, their potential for high-quality natural language understanding and generation is matched by the technical difficulties of serving them in production environments. Below is a deeper exploration of the key challenges and emerging solutions.

Latency in Inference

Latency remains one of the most critical challenges in LLM inference, especially for real-time applications. Unlike traditional models, which can often process entire inputs in parallel, LLMs use autoregressive decoding during text generation. This means they generate outputs token by token, where the computation of each token depends on the previous one. This sequential nature introduces inherent delays, particularly noticeable in tasks requiring long-form outputs, such as content generation or summarization.

Further exacerbating latency issues is the variability in request complexity. Some inputs may require only a few steps of computation, while others—due to higher token counts or more complex prompts—demand significantly longer processing times. Balancing these requirements while maintaining consistent response times is an ongoing area of optimization.

Computational Demands and Resource Constraints

LLMs are computationally intensive. A single inference operation can require trillions of floating-point operations (FLOPs), even for moderately sized inputs. These demands necessitate the use of high-performance hardware, such as GPUs or TPUs. However, such hardware is expensive and limited in availability, making large-scale deployment a costly endeavor.

Moreover, memory requirements for LLM inference are immense. For instance, storing model weights for a 175-billion-parameter model like GPT-3 requires over 700 GB of memory in its full-precision form. This memory requirement grows when considering additional overhead for processing large batch sizes, caching intermediate computations, or handling multiple concurrent requests. Techniques like model quantization, weight sharing, and offloading parts of the computation to disk or slower memory are frequently used to mitigate this challenge but often at the expense of throughput or accuracy.

Scalability and Multitenancy

Scalability is essential for deploying LLMs in environments with high and variable traffic. Inference systems must handle thousands or even millions of concurrent requests while ensuring consistent quality and low latency. This challenge becomes more pronounced in multitenant systems, where multiple users or applications share the same underlying infrastructure. Resource allocation in such environments must be dynamic and efficient to avoid resource contention or overprovisioning.

Load balancing is a critical component of scalability. Requests must be distributed intelligently across available hardware to ensure that no single device becomes a bottleneck. Strategies such as request-level load balancing, horizontal scaling (replicating the model across devices), and vertical scaling (improving individual device performance) are common solutions, though they introduce their own complexities in deployment and maintenance.

The Trade-Off Between Batching and Responsiveness

Batching multiple inference requests is a widely used technique to improve hardware utilization and throughput. By grouping requests, the system can process them in parallel, leveraging the full computational capabilities of GPUs or TPUs. However, batching comes with a trade-off: as batch sizes increase, the time individual requests spend waiting for others to join the batch grows, leading to higher latency.

Dynamic batching algorithms aim to strike a balance between these competing goals. By adaptively adjusting batch sizes based on workload and latency requirements, these systems can optimize for both throughput and responsiveness. Nonetheless, fine-tuning these algorithms is complex and often requires a deep understanding of both hardware and application-specific requirements.

Model Parallelism and Distributed Systems

For extremely large models, it is often impossible to fit the entire model into the memory of a single device. Model parallelism, where the model is split across multiple devices, is a common solution. However, this approach introduces communication overhead, as devices need to exchange data during inference. Latency and bandwidth constraints in distributed systems can become bottlenecks, particularly when deploying across geographically distributed data centers.

Pipeline parallelism, which segments the model into stages processed in a pipeline fashion, can alleviate some of these issues but requires careful scheduling and synchronization to avoid idle devices. Combining pipeline parallelism with other techniques, such as tensor parallelism (splitting computations across devices), can yield further optimizations but adds to the complexity of implementation.

Cost Efficiency

The financial cost of LLM inference is another major concern. Deploying a single large model at scale can lead to significant expenses in hardware acquisition, energy consumption, and operational maintenance. For businesses, these costs can quickly become prohibitive, especially if the model is used in applications with low profit margins.

One emerging approach to cost efficiency is using smaller, distilled versions of large models for inference. Knowledge distillation transfers the knowledge from a large "teacher" model to a smaller "student" model, which can approximate the teacher's performance while being faster and cheaper to run. Similarly, serverless architectures and spot instances are being explored to dynamically scale infrastructure costs based on demand.

Reliability and Robustness

Inference systems must be not only fast and scalable but also reliable and robust. Ensuring that an LLM produces consistent and accurate results under varying conditions is a persistent challenge. For example, minor variations in input phrasing can sometimes lead to drastically different outputs. Furthermore, system failures, such as hardware outages or network delays, can disrupt service quality.

Robust monitoring and failover mechanisms are essential to mitigate these risks. Techniques like request retries, checkpointing, and fallback models (smaller models that can serve as a backup) are often employed to ensure reliability. Additionally, fine-tuning models for specific tasks or domains can enhance robustness by reducing output variability and improving contextual understanding.

Ethical and Security Considerations

Inference systems for LLMs are not immune to ethical and security challenges. Outputs must be monitored to avoid generating harmful or biased content. Real-time filtering mechanisms or safety layers can help mitigate these risks but add additional computational overhead.

Moreover, deploying LLMs as APIs or services exposes them to potential abuse, such as adversarial inputs designed to exploit the model or denial-of-service (DoS) attacks targeting the infrastructure. Security measures, including input validation, rate limiting, and anomaly detection, are critical to maintaining the integrity and reliability of these systems.

LLM Memory Optimization

Memory optimization is an essential focus in the deployment and training of large language models (LLMs). These models, with their immense parameter sizes and resource requirements, often push the limits of modern hardware. To make them practical and scalable for real-world applications, researchers and engineers have developed sophisticated techniques to optimize memory usage. These optimizations span hardware, software, and algorithmic domains, addressing challenges that arise in both training and inference contexts.

The Memory Challenges of LLMs

At the heart of LLM memory usage are three main components: model weights, activations, and gradients. Model weights are the parameters learned during training and used during inference, while activations are the intermediate results generated during computation. Gradients, relevant during training and fine-tuning, represent the derivatives used to update the weights.

Each of these components requires significant memory, and their combined requirements can exceed the capabilities of even high-end GPUs or TPUs. For instance:

- A single layer in a transformer model might have billions of parameters, and models with hundreds of layers are now common.
- Activations scale with both the number of parameters and the input sequence length, particularly in attention mechanisms, where the memory scales quadratically with the sequence length.

- Gradients require storage of additional memory copies of weights and activations during backpropagation.

Given these demands, optimizing memory usage is a critical step in ensuring the viability of LLMs across various applications.

Key Memory Optimization Techniques

1. **Precision Reduction (Quantization)**

   One of the most effective methods for memory optimization is reducing the numerical precision of model weights and activations. Models typically operate in 32-bit floating-point (FP32) precision during training. Reducing this to FP16 (half-precision) or INT8 (integer precision) can halve or even quarter the memory footprint. Advances in quantization-aware training allow models to retain nearly the same performance even at reduced precision. Some techniques dynamically adjust precision during computation to maintain critical details while optimizing memory.

2. **Gradient Checkpointing (Activation Recomputation)**

   In a standard training process, activations from the forward pass are stored for use during backpropagation. For very large models, this storage becomes a memory bottleneck. Gradient checkpointing addresses this by saving only a subset of activations during the forward pass and recomputing them as needed during backpropagation. While this approach increases computation time, it drastically reduces memory usage, enabling larger models to be trained on the same hardware.

3. **Model Offloading**

   Offloading involves moving parts of the model or activations from GPU memory to CPU memory or even disk storage. This technique takes advantage of the larger capacity of slower storage mediums to hold less frequently accessed data. For example, weights of layers that are not currently being used can be temporarily offloaded and loaded back when needed. Advances in memory management algorithms ensure minimal latency in retrieving offloaded data, making this approach viable for both training and inference.

4. **Optimized Attention Mechanisms**

   The attention mechanism, a cornerstone of transformer-based LLMs, is one of the largest consumers of memory, scaling quadratically with the input sequence length. Techniques such as sparse attention, sliding window attention, and low-rank approximations have been developed to reduce the memory requirements of attention computations. These methods approximate the full attention mechanism while maintaining high accuracy, significantly cutting memory usage for long sequences.

5. **Model Parallelism**

   When a model is too large to fit into a single device, model parallelism splits the model across multiple devices. In tensor parallelism, individual layers are divided across devices, with computations performed in parallel. Pipeline parallelism further splits the model into stages that run in sequence but across different devices. Both approaches distribute memory usage but introduce challenges such as communication overhead and synchronization, which must be carefully managed to avoid bottlenecks.

6. **Memory-Efficient Architectures**

   Designing architectures with memory efficiency in mind is another approach. Techniques such as reversible layers, where intermediate activations can be reconstructed instead of stored, reduce memory requirements during both training and inference. Some emerging architectures are explicitly designed to minimize memory usage while maintaining the expressive power of traditional transformers.

7. **Compression Techniques (Pruning and Distillation)**

   Pruning removes redundant parameters from the model, reducing the size of the model without significantly impacting performance. For example, sparsity can be introduced by identifying weights that contribute minimally to outputs and setting them to zero. Knowledge distillation takes this a step further by training a smaller "student" model to replicate the behavior of a larger "teacher" model. The result is a more compact model with lower memory requirements, ideal for inference on resource-constrained devices.

8. **Dynamic and Adaptive Batching**

   During inference, batching multiple requests together improves efficiency but increases memory usage. Dynamic batching algorithms analyze the available memory and workload in real time, adjusting batch sizes accordingly. Micro-batching, where a large batch is split into smaller sub-batches processed sequentially, ensures that memory constraints are respected without sacrificing throughput.

9. **Unified Memory Architectures**

   Modern hardware advancements, such as unified memory architectures, allow models to utilize both high-speed GPU memory and larger, slower system memory seamlessly. This hierarchical memory management ensures frequently accessed data remains in faster memory, reducing bottlenecks caused by offloading.

Trade-Offs in Memory Optimization

While these techniques can significantly reduce memory requirements, they often come with trade-offs:

- **Computation Time**: Techniques like gradient checkpointing and offloading save memory but increase computational overhead, leading to longer training or inference times.

- **Accuracy**: Quantization and pruning, while reducing memory, may lead to small losses in model performance, requiring careful tuning.
- **Complexity**: Implementing advanced memory optimization techniques, such as model parallelism or custom attention mechanisms, adds complexity to system design and maintenance.

Future Directions

As LLMs continue to grow in size and importance, memory optimization will remain a critical area of research and innovation. Some emerging trends include

- **Hardware-Specific Optimizations**: New hardware, such as custom AI accelerators (e.g., NVIDIA's Hopper GPUs or Google's TPUv5), is being designed with memory optimization in mind, providing native support for techniques like quantization and memory-efficient attention.
- **Neurosymbolic Systems**: Combining neural models with symbolic reasoning systems can reduce memory usage by offloading some tasks to more efficient symbolic systems.
- **Federated and Decentralized Models**: Splitting computations across distributed devices or edge systems can alleviate memory bottlenecks, particularly for real-time applications.

Memory optimization is a cornerstone of making LLMs scalable, accessible, and efficient. By addressing memory constraints through a combination of hardware advances, algorithmic innovations, and architectural adjustments, the transformative potential of LLMs can be realized in a wide array of applications, from consumer devices to enterprise systems.

LLM Compression

Compression techniques for large language models (LLMs) are essential to address the challenges posed by their massive size and computational demands. These models often contain hundreds of billions of parameters, resulting in substantial memory requirements and high inference costs. Compression aims to reduce the model's size and computational complexity while preserving its performance, making it feasible to deploy LLMs in resource-constrained environments or at scale.

The need for compression arises because the size of LLMs directly impacts their latency, energy consumption, and cost of deployment. Without compression, the operational requirements of LLMs are prohibitive for many real-world applications, especially for edge devices or real-time systems. Effective compression strikes a balance between model size and performance, ensuring that accuracy and generalization are retained even as the model is scaled down.

Quantization

Quantization is one of the most widely used compression techniques for LLMs. It involves reducing the numerical precision of the model's weights and activations. For example, full-precision 32-bit floating-point (FP32) representations can be converted to 16-bit (FP16) or 8-bit integers (INT8). This reduction significantly decreases the memory footprint and computational overhead of the model.

Quantization can be applied in different stages of model deployment. During training, quantization-aware training ensures that the model learns to operate effectively at lower precisions. Post-training quantization, applied after the model is trained, is simpler to implement but may result in minor accuracy degradation. Advances in this area, such as mixed-precision quantization and adaptive quantization, allow further optimization by using lower precision for less critical parts of the model while retaining higher precision for sensitive components.

Pruning

Pruning reduces a model's size by identifying and removing parameters that contribute minimally to its performance. This approach assumes that many of the parameters in LLMs are redundant and can be safely eliminated without significantly affecting accuracy.

There are several methods for pruning, including structured pruning, which removes entire layers, filters, or attention heads, and unstructured pruning, which targets individual weights. Pruning is often iterative: the model is pruned and then fine-tuned to recover any lost performance. While structured pruning results in models that are easier to implement on hardware, unstructured pruning often achieves higher compression ratios at the cost of increased deployment complexity.

Knowledge Distillation

Knowledge distillation trains a smaller "student" model to mimic the behavior of a larger "teacher" model. The student model learns not only from the teacher's outputs but also from the intermediate representations and logits generated by the teacher during training. This process transfers knowledge from the larger model to the smaller one, enabling the student model to achieve similar performance with significantly fewer parameters.

Distillation is particularly effective for compressing LLMs while retaining their accuracy. It is widely used in scenarios where the smaller model must operate in latency-sensitive environments, such as mobile devices or edge computing. The resulting student models are faster and more memory-efficient, making them suitable for deployment without significant hardware investments.

Low-Rank Factorization

Low-rank factorization is a mathematical approach that approximates the large weight matrices of LLMs with smaller, low-rank matrices. Since many of the learned parameters in neural networks are redundant, this technique leverages the inherent structure of these matrices to reduce their size.

By decomposing weight matrices into smaller components, low-rank factorization can reduce memory requirements and computational complexity. This method is particularly effective for compressing fully connected layers and attention mechanisms, which often dominate the size of LLMs.

Sparsity-Inducing Techniques

Sparsity-inducing techniques aim to make LLMs more efficient by introducing sparsity into their parameters or activations. Sparse models only activate or utilize a subset of their weights for any given input, significantly reducing computation and memory usage.

Techniques like sparse attention mechanisms focus on reducing the quadratic complexity of traditional attention by limiting computations to relevant portions of the input. Similarly, sparsity in weight matrices can be achieved through training with regularization techniques like L1 or L2 penalties or by applying threshold-based pruning during or after training.

Compression Challenges and Trade-Offs

While compression significantly reduces the size and computational demands of LLMs, it comes with trade-offs. Reducing precision or pruning weights may lead to slight degradation in model accuracy, particularly for tasks requiring nuanced understanding or generation. Knowledge distillation, while effective, requires additional training cycles, increasing the computational cost during the compression phase.

Another challenge lies in the implementation of compressed models on hardware. Techniques like pruning or sparsity require specialized software and hardware optimizations to fully realize their benefits. For example, unstructured sparsity may lead to irregular memory access patterns, reducing efficiency on standard GPUs or CPUs. As a result, the choice of compression techniques often depends on the target deployment environment and available hardware capabilities.

Future Directions in LLM Compression

Advances in LLM compression continue to evolve as researchers aim to balance performance, size, and efficiency. Techniques such as dynamic pruning, which adjusts model size based on input complexity, and hybrid methods that combine quantization with pruning or distillation are gaining attention. Additionally, hardware innovations, such as custom accelerators designed to handle compressed models, are making it easier to deploy LLMs in resource-constrained settings.

Another emerging trend is task-specific compression, where a general-purpose LLM is fine-tuned and compressed for specific applications. This approach allows the model to retain high performance on targeted tasks while reducing its size and resource requirements.

In conclusion, LLM compression is a critical area of research and practice that enables the deployment of these powerful models in diverse environments. By employing techniques like quantization, pruning, knowledge distillation, and low-rank factorization, organizations can make LLMs more efficient and accessible, unlocking their potential in a wider range of applications. As the demand for scalable and efficient AI systems grows, innovations in compression will play a central role in shaping the future of LLM deployment.

Attention Layer Optimization

The attention mechanism, particularly in transformer-based architectures, is a foundational component of large language models (LLMs). It enables models to identify and focus on relevant parts of the input sequence, capturing dependencies between tokens regardless of their distance from one another. However, the attention mechanism is also one of the most resource-intensive components of these models, with its memory and computational costs scaling quadratically with the input sequence length. This presents significant challenges in both training and inference, especially for tasks involving long documents or real-time processing. Attention layer optimization seeks to address these challenges by improving the efficiency of this mechanism while maintaining or enhancing its performance.

The quadratic complexity of standard self-attention arises from the need to compute attention scores for all pairs of tokens in the input sequence. For an input sequence of length nnn, this requires O(n2)O(n^2)O(n2) operations and memory, which becomes impractical for large nnn. This limitation drives the need for optimizations that reduce the computational and memory overhead of attention layers without sacrificing the quality of the model's outputs.

One approach to optimization is the use of **sparse attention mechanisms**. Unlike dense attention, which calculates scores for every pair of tokens, sparse attention restricts the computation to a subset of token pairs based on predefined patterns or learned relevance. For example, sliding window attention only considers a fixed number of neighboring tokens for each position, significantly reducing the computational burden. Similarly, global-local attention mechanisms combine local attention for nearby tokens with global attention for a few critical tokens, striking a balance between efficiency and expressiveness.

Another method involves **low-rank approximations**, which approximate the attention matrix using techniques like singular value decomposition (SVD) or low-rank factorization. These methods exploit the observation that attention matrices often have low intrinsic rank, meaning much of the information can be captured using a smaller number of components. By reducing the dimensionality of the attention computation, low-rank approximations reduce both memory usage and computational requirements.

For applications involving very long sequences, **hierarchical attention** mechanisms have proven effective. In this approach, the model processes the input in chunks, computing attention within each chunk before aggregating information across chunks. This hierarchical structure reduces the number of pairwise comparisons required, enabling the processing of much longer sequences without incurring prohibitive costs.

Efficient attention kernels have also been developed to leverage hardware-specific optimizations. These kernels are tailored for parallel computation on GPUs and TPUs, minimizing memory access bottlenecks and maximizing throughput. For instance, some implementations use fused operations that combine multiple computation steps into a single kernel call, reducing overhead and improving efficiency.

Another avenue for optimization is the incorporation of **approximate algorithms** that simplify the attention computation. For example, random feature methods approximate the softmax function used in attention calculations, enabling linear rather than quadratic scaling. These methods introduce minor approximations to the final results but significantly accelerate computation, making them suitable for latency-sensitive applications.

Optimizing attention layers also involves modifying the model's architecture to be more efficient. Techniques like reformer models and performers replace standard attention mechanisms with alternative formulations that are inherently more scalable. These models achieve linear or near-linear complexity in terms of sequence length, making them practical for processing very large inputs.

Despite these advancements, attention layer optimization is not without trade-offs. Reducing the computational and memory requirements often involves approximations or simplifications that can degrade model performance, especially for tasks requiring fine-grained or global contextual understanding. Therefore, the choice of optimization technique depends on the specific application and its requirements for accuracy, latency, and resource availability.

Looking forward, the development of hybrid approaches that combine multiple optimization techniques is a promising area of research. For instance, combining sparse attention with low-rank approximations or hierarchical attention with efficient kernels can yield even greater efficiency gains. Furthermore, advances in hardware design, such as specialized AI accelerators, are expected to further enhance the practicality of optimized attention mechanisms.

In summary, attention layer optimization is critical for making LLMs scalable and efficient. By reducing the computational and memory demands of the attention mechanism, these optimizations enable the deployment of LLMs in a broader range of applications, from real-time systems to tasks involving extremely long documents. As the complexity and utility of LLMs continue to grow, innovations in attention layer optimization will remain a central focus in the evolution of AI architectures.

Scheduling Optimization in LLM Deployment

Scheduling optimization is a vital aspect of deploying large language models (LLMs), ensuring efficient allocation of computational resources to meet the diverse demands of real-world applications. LLM inference is a resource-intensive process, requiring significant compute and memory, often under stringent latency constraints. Scheduling optimization involves orchestrating tasks, allocating hardware resources, and managing workloads to maximize throughput, minimize latency, and balance system utilization.

The complexity of scheduling arises from the variability in LLM workloads. Input sizes, model architectures, and user demands can differ significantly, making static scheduling strategies inefficient. Effective scheduling optimization dynamically adjusts to these variations, enabling the deployment of LLMs in environments ranging from high-throughput server clusters to latency-critical edge devices.

One of the foundational challenges in scheduling optimization is balancing **batching** and **responsiveness**. Batching groups multiple requests into a single computation to maximize hardware utilization, as modern accelerators like GPUs and TPUs perform more efficiently with larger workloads. However, batching can introduce delays for individual requests, particularly in latency-sensitive applications such as chatbots or virtual assistants. Dynamic batching algorithms address this trade-off by adaptively adjusting batch sizes based on current workloads and system conditions, ensuring a balance between efficiency and responsiveness.

- **Request-level scheduling** focuses on managing individual inference requests in a way that meets application-specific requirements. For instance, in a multitenant environment, different applications may have varying latency and accuracy priorities. Scheduling strategies must allocate resources accordingly, ensuring that high-priority tasks are not delayed by lower-priority workloads. This often involves implementing sophisticated priority queues, resource allocation policies, and preemption mechanisms.

- **Batch-level scheduling** expands this concept to aggregate workloads. It determines how requests are grouped into batches and assigns these batches to available hardware. The goal is to maximize hardware utilization without exceeding memory limits or causing contention among processes. Efficient batch-level scheduling often relies on predictive algorithms that anticipate workloads based on historical patterns or incoming request rates, allowing the system to preemptively allocate resources and adjust batch sizes.
- **Iteration-level scheduling** comes into play during training or iterative inference processes, such as fine-tuning or beam search. These processes involve multiple steps, each with varying resource requirements and dependencies. Effective scheduling ensures that the necessary resources are available at each step, minimizing idle time and synchronization delays. For distributed training setups, iteration-level scheduling must also account for interdevice communication, ensuring that data transfers are efficiently managed to prevent bottlenecks.
- **Continuous batching** is a dynamic approach that handles incoming requests on a rolling basis, rather than waiting for a fixed batch size or time window. This technique is particularly useful for real-time systems where input patterns are unpredictable. By continuously adjusting the batch size and processing intervals based on the current system state, continuous batching minimizes latency while maintaining high throughput.

The underlying hardware architecture plays a significant role in scheduling optimization. Modern accelerators offer features like multistream processing and hardware virtualization, enabling concurrent execution of multiple tasks. Scheduling algorithms must leverage these capabilities effectively, distributing workloads to maximize parallelism and minimize contention. Additionally, heterogeneity in hardware resources, such as a mix of CPUs, GPUs, and TPUs, introduces another layer of complexity, requiring intelligent scheduling strategies that assign tasks to the most suitable device based on task characteristics and hardware capabilities.

Communication and synchronization in distributed systems also affect scheduling optimization. In scenarios where models are split across multiple devices (e.g., model parallelism or pipeline parallelism), scheduling must account for data dependencies and interdevice communication overhead. Techniques like overlapping computation with communication, scheduling communication-intensive tasks during idle periods, and optimizing data transfer paths are crucial for maintaining efficiency in distributed setups.

Scheduling optimization must also consider **energy efficiency** and **cost constraints**, especially in cloud environments. Dynamically scaling resources based on demand, leveraging spot instances, and utilizing energy-aware scheduling algorithms can reduce operational costs while maintaining service quality. For edge deployments, where energy and compute resources are limited, scheduling strategies must minimize resource usage without compromising performance.

Finally, scheduling optimization is increasingly incorporating **machine learning-driven approaches**. Predictive models trained on historical data can forecast workload patterns, enabling proactive resource allocation and batch adjustments. Reinforcement learning algorithms can dynamically adapt scheduling policies based on real-time feedback, continuously improving efficiency over time.

In summary, scheduling optimization in LLM deployment is a multifaceted challenge that balances efficiency, responsiveness, and cost. By orchestrating tasks and resources effectively across various levels—request, batch, and iteration—scheduling ensures that LLMs can meet the demands of diverse applications. As LLMs continue to grow in size and complexity, advances in scheduling strategies will play a critical role in enabling scalable, cost-effective, and high-performance deployments.

Summary

This chapter offers a comprehensive and practical guide to deploying LLM-powered applications, bridging the gap between cutting-edge AI models and real-world usability. It excels in explaining the technical complexities of deployment, covering everything from cloud infrastructure and optimization strategies to scheduling and memory management.

By clearly outlining three integration pathways—hosted, prepackaged, and open source—the chapter empowers readers to choose an approach aligned with their technical expertise, privacy needs, and cost considerations. It also dives deep into advanced topics like attention layer optimization, model compression, and scalability techniques, providing an invaluable toolkit for practitioners.

Overall, this chapter is an essential resource for anyone looking to operationalize LLMs efficiently and responsibly, balancing performance, scalability, and ethics.

CHAPTER 5

Building and Fine-Tuning LLMs

The transformative power of large language models (LLMs) has reshaped industries, redefined human-computer interaction, and expanded the boundaries of artificial intelligence. As these models grow in size and sophistication, so too does the complexity of building and fine-tuning them. This chapter delves into the art and science of developing LLMs, providing a road map for practitioners seeking to navigate the intricacies of this fascinating domain.

LLMs, such as GPT, BERT, and their derivatives, are pretrained on a vast corpora, enabling them to perform a range of tasks, from text generation to sentiment analysis and beyond. However, harnessing the full potential of these models often requires customizing them for specific applications. **This is where fine-tuning comes into play**. Fine-tuning not only optimizes the model for domain-specific use cases but also enhances its efficiency and effectiveness by tailoring it to unique datasets and requirements.

In this chapter, we explore both the foundational principles and practical techniques of building and fine-tuning LLMs. Whether you are an AI researcher, a data scientist, or a developer, understanding these principles is key to unlocking the potential of LLMs for your projects.

Key Themes of This Chapter

- **Understanding the Foundations of LLMs:** We begin by examining the architectural components that make LLMs so powerful. From transformers to attention mechanisms, we unravel the building blocks that enable these models to achieve remarkable feats in natural language understanding and generation.

D. Grigorov, *Intermediate Python and Large Language Models*, https://doi.org/10.1007/979-8-8688-1475-4_5

- **The Pretraining Paradigm:** Pretraining is the backbone of LLMs. By training on diverse and extensive datasets, these models learn generalizable patterns and relationships within language. We'll discuss how pretraining is conducted and its impact on downstream applications.
- **Fine-Tuning Strategies:** Fine-tuning transforms a general-purpose LLM into a task-specific powerhouse. We'll walk through different fine-tuning methodologies, including supervised fine-tuning, instruction tuning, and reinforcement learning from human feedback (RLHF).
- **Practical Considerations and Challenges:** Building and fine-tuning LLMs come with unique challenges, from computational requirements to ethical considerations. We provide insights into overcoming these hurdles and ensuring responsible AI deployment.

The rapid evolution of LLMs means that staying current with techniques for building and fine-tuning them is more critical than ever. The ability to adapt these models to meet specific needs is what distinguishes a good implementation from a transformative one. Moreover, as ethical concerns and biases in AI take center stage, it is imperative to approach the fine-tuning process with responsibility and care.

By the end of this chapter, you will have a comprehensive understanding of how to build and fine-tune LLMs, equipping you with the knowledge to bring cutting-edge AI solutions to life. Whether you aim to improve customer experiences, automate complex workflows, or pioneer new frontiers in AI, the principles outlined here will serve as your guide.

Embark on this journey to unravel the intricacies of LLMs and discover how to harness their immense potential for innovation and impact.

Architecture of Large Language Models (LLMs)

Large language models (LLMs), such as GPT-4 and BERT, are intricate systems designed to process, comprehend, and generate humanlike text. These models are powered by the Transformer architecture, a revolutionary framework that enables them to capture the complexities of language and context. Through multiple interconnected layers and components, LLMs achieve their remarkable capabilities in tasks ranging from text generation to translation and beyond.

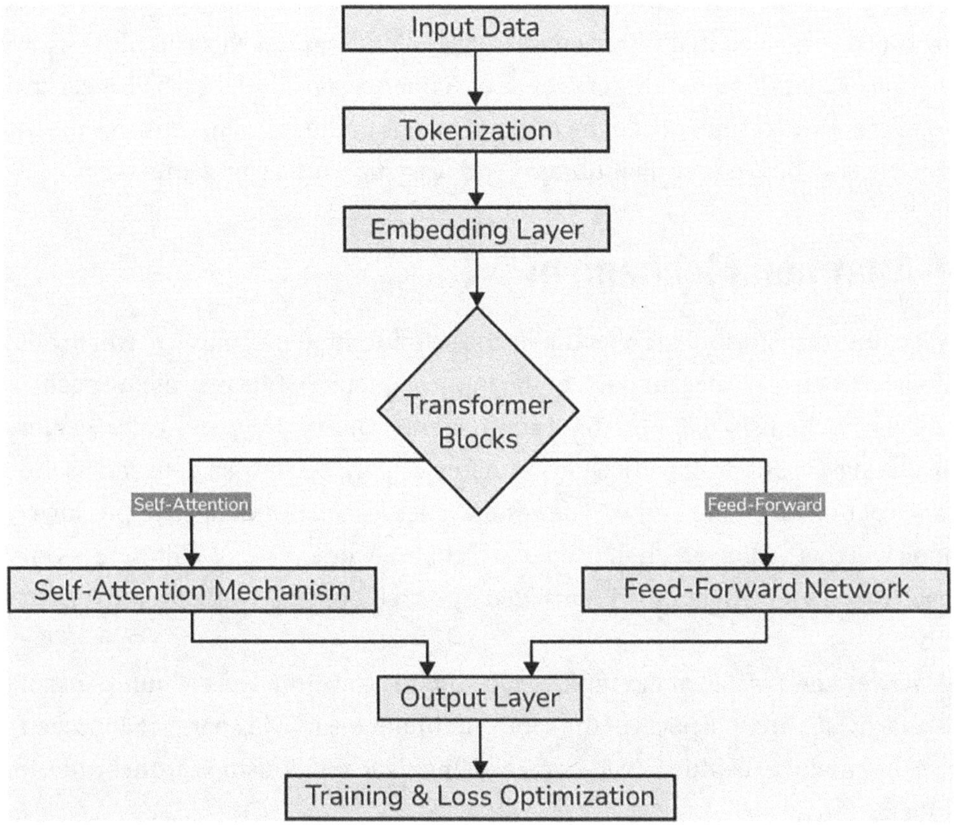

Figure 5-1. *Large Language Models Common Architecture*

At the Foundation of Any LLM Lies the Process of Tokenization

This is where the input text is divided into smaller units called tokens, which could be entire words, subwords, or even individual characters. Tokenization helps the model handle text efficiently, particularly in cases involving rare or compound words. After this step, each token is transformed into a numerical representation or embedding that the model can process.

The embedding layer plays a critical role in translating tokens into dense vectors in a high-dimensional space. These vectors capture the semantic and syntactic meanings of the tokens, enabling the model to understand their relationships. Word embeddings can either be pretrained, as in techniques like Word2Vec and GloVe, or learned dynamically during training, as seen in modern contextual embeddings like those in BERT. However, since Transformers lack an inherent sense of sequence, positional embeddings are added to these vectors to encode the order of tokens in the text. This ensures the model can differentiate between similar phrases with varying word arrangements.

Self-Attention Mechanism

Central to the Transformer architecture is the self-attention mechanism, which allows the model to focus on relevant parts of the input sequence while processing each token. This mechanism relies on three components: **Query, Key, and Value vectors.** By calculating the dot product of Query and Key vectors, the model determines the relevance of one token to another. The resulting scores are normalized to produce attention weights, which are then used to weight the Value vectors. This process ensures that each token's representation is enriched by its relationship with other tokens in the sequence.

Moreover, the Transformer employs **multihead attention**, where multiple attention heads analyze different aspects of the input simultaneously. This parallel approach enables the model to capture a wide range of linguistic relationships, from syntax to semantics.

Once the self-attention mechanism completes its operation, the output is passed through a **feedforward neural network**. This component consists of fully connected layers with nonlinear activation functions, such as **ReLU,** which allow the model to learn complex transformations. The output of the feedforward network refines the **token representations, enabling the model to discern intricate patterns in the data.**

Layer Normalization and Residual Connections

To stabilize the training process and improve gradient flow, the architecture incorporates layer normalization and residual connections.

- Layer normalization ensures consistent input scaling across layers, reducing the risk of vanishing or exploding gradients.
- Residual connections, on the other hand, add the input of a layer directly to its output, preserving critical information and allowing for the successful training of deeper networks.

Transformer Blocks

The architecture stacks multiple Transformer blocks, each containing self-attention and feedforward layers, to learn hierarchical representations of the input. Early layers in the stack capture surface-level features like word boundaries, while deeper layers focus on abstract, semantic relationships. This hierarchical approach enables LLMs to process text at various levels of complexity, from syntax to context and meaning.

At the Output Layer, LLMs Operate Differently Depending on Their Design

In **autoregressive models like GPT**, the objective is to predict the next token in a sequence based on the tokens that precede it. In masked language models like BERT, the model is trained to predict missing or masked tokens, leveraging the context on both sides of the sequence.

Regardless of the objective, the final step involves passing the output through a **softmax layer,** which converts the model's predictions into probabilities over the vocabulary. The most probable token is then selected, completing the model's task of generating or understanding text.

This sophisticated architecture enables LLMs to perform a wide range of natural language processing tasks with exceptional accuracy and fluency. By combining innovations like self-attention, hierarchical representations, and efficient embedding techniques, these models have transformed how we interact with and leverage language in technology. From powering chatbots to advancing scientific research, LLMs continue to redefine the possibilities of artificial intelligence.

Variations in LLM Architectures

LLMs built on the Transformer architecture have evolved into three main categories, each optimized for specific tasks. Below is a detailed look at these categories:

1. **Autoencoders**
    - **Definition**: Use only the encoder part of the Transformer architecture while omitting the decoder after pretraining.
    - **Examples**: Models like BERT (Bidirectional Encoder Representations from Transformers) and RoBERTa.
    - **Use Cases**: Ideal for tasks requiring understanding of context, such as sentiment analysis, text classification, and named entity recognition.
    - **Training Methodology**: Trained using Masked Language Modeling (MLM), where specific words or tokens in a sequence are masked, and the model learns to predict them. This approach enhances the model's contextual understanding.
2. **Auto-Regressors**
    - **Definition**: Use the decoder part of the Transformer while discarding the encoder after pretraining.
    - **Examples**: GPT series (Generative Pretrained Transformer) and BLOOM.
    - **Use Cases**: Designed for text generation, story writing, question answering, and summarization. These models excel in generating coherent and contextually relevant text.
    - **Training Methodology**: Employ Causal Language Modeling, where the model predicts the next token in a sequence based on preceding tokens, allowing it to generate sequential outputs.
3. **Sequence-to-Sequence Models**
    - **Definition**: Incorporate both the encoder and decoder components of the Transformer.

- **Examples**: Models like T5 (Text-to-Text Transfer Transformer) and BART (Bidirectional and Auto-Regressive Transformers).
- **Use Cases**: Versatile models suited for translation, summarization, and question-answering tasks.
- **Training Methodology**: Often trained with techniques like span corruption, where parts of the input are deliberately distorted, and the model learns to reconstruct the original.

Fine-Tuning Strategies and Considerations

What Is LLM Fine-Tuning?

Fine-tuning involves adjusting the parameters of a pretrained large language model to better suit a specific task or domain. While models like GPT are equipped with extensive general language knowledge, they lack the specialized expertise required for certain areas. Fine-tuning overcomes this limitation by enabling the model to learn from domain-specific data, enhancing its accuracy and effectiveness for particular applications.

Through fine-tuning, the model is exposed to task-specific examples, allowing it to grasp the subtleties and nuances of the domain. This process transforms a general-purpose language model into a specialized tool, maximizing the potential of LLMs for targeted use cases.

Fine-tuning is particularly useful in scenarios where you need the following:

Customization

Every domain or task comes with its own distinct language patterns, terminologies, and contextual intricacies. Fine-tuning a pretrained LLM enables customization, allowing the model to better understand these unique characteristics and generate domain-specific content. This tailoring ensures that the model's outputs align with your specific requirements, delivering accurate and contextually relevant results.

Whether you are working with legal documents, medical records, business reports, or proprietary company data, fine-tuning empowers LLMs to provide specialized expertise. By training the model on domain-specific datasets, you can harness the capabilities of LLMs while ensuring they meet the precision and relevance demanded by your use case.

Data Compliance

Industries like healthcare, finance, and law operate under stringent regulations governing the use and protection of sensitive information. Fine-tuning LLMs on proprietary or regulated data allows organizations to develop models that comply with data privacy and security standards.

This approach minimizes the risks of exposing sensitive information to external systems while creating models that are securely trained on in-house or industry-specific data. Fine-tuning enhances the privacy, security, and regulatory compliance of LLM applications.

Limited Labeled Data

In many practical scenarios, acquiring large volumes of labeled data for specific tasks or domains is both challenging and costly. Fine-tuning addresses this issue by making the most of existing labeled data, enabling a pretrained LLM to adapt effectively to smaller datasets.

This method allows organizations to overcome data scarcity while still achieving notable improvements in the model's performance and relevance. Even with limited labeled data, fine-tuning ensures the model delivers accurate and reliable results tailored to the task or domain.

Data Requirements for Fine-Tuning

To fine-tune a large language model (LLM) effectively, it's critical to understand the data requirements necessary to support both training and validation. Below are key guidelines to ensure a successful fine-tuning process:

1. **Use a Large Dataset**

   The size of the training and validation dataset should align with the complexity of the task and the model being fine-tuned. Typically, thousands or tens of thousands of examples are recommended. While larger models can learn more effectively from smaller datasets, having sufficient data is still essential to prevent overfitting or eroding the knowledge gained during the pretraining phase.

2. **Ensure High Data Quality**

   The dataset should be clean, consistent, and free from incomplete or incorrect examples. High-quality data helps the model learn effectively and reduces the risk of introducing errors or biases during fine-tuning.

3. **Use a Representative Dataset**

   The fine-tuning dataset should accurately reflect the types of data the model will encounter in its intended use. For example, if fine-tuning for sentiment analysis, the dataset should include data from diverse sources, genres, and domains, capturing the range of human emotions. Balanced distribution across categories (e.g., positive, negative, neutral sentiments) is also important to prevent skewed predictions.

4. **Provide Sufficiently Specified Inputs**

   The dataset should contain clear and detailed input information to guide the desired output. For instance, when fine-tuning a model for email generation, inputs should include specific prompts that direct the model's creativity and relevance. Additionally, the dataset should define expectations for length, style, and tone, ensuring that the model generates outputs aligned with your requirements.

LLM Fine-Tuning Techniques

Primary Approaches to Fine-Tuning

Finc-tuning large language models (LLMs) is the process of adjusting their parameters to meet specific task requirements. The extent of these adjustments depends on the complexity of the task and the desired outcome. Broadly, two primary approaches to fine-tuning have emerged: **feature extraction** and **full fine-tuning**. Each method offers unique strengths and trade-offs. Let's explore them.

a. Feature Extraction (Repurposing)

Feature extraction, often referred to as repurposing, treats the pretrained LLM as a fixed feature extractor. This approach capitalizes on the model's vast knowledge, which has been developed by training on expansive datasets covering a variety of language features and patterns.

In this method, the majority of the model remains frozen, while only the final layers are trained on task-specific data. By focusing adjustments on these layers, the model adapts its rich, pre-existing representations to the specific requirements of the task. This approach is particularly efficient, as it minimizes computational costs and training time while still delivering reliable, domain-specific results. Feature extraction is ideal for tasks where the pretrained model's foundational understanding suffices and only minor refinements are needed.

b. Fine-Tuning Embedding Models

Fine-tuning embedding models for large language models (LLMs) is a powerful technique to adapt pretrained models to specific tasks or domains, improving their performance on downstream applications like text classification, semantic search, question answering, or clustering. Embedding models, which convert text into dense vector representations, are a core component of LLMs, capturing semantic and syntactic relationships between words, phrases, or entire documents.

Embedding models in LLMs (e.g., BERT, RoBERTa, or sentence-transformers) map input text into a high-dimensional vector space where similar meanings are positioned closer together. Pretrained LLMs come with embeddings learned from vast, general-purpose datasets, but these embeddings may not be optimal for specialized tasks or domains (e.g., medical texts, legal documents, or informal social media language). Fine-tuning adjusts these embeddings to better align with the target task or data.

Why Fine-Tune Embedding Models?

- **Domain Adaptation:** Pretrained embeddings may not capture domain-specific nuances.
- **Task-Specific Optimization:** Fine-tuning tailors embeddings to prioritize features relevant to a specific task, like sentiment analysis or entity recognition.

- **Improved Performance:** Adjusted embeddings often lead to better accuracy, precision, or recall in downstream applications.
- **Efficiency:** Fine-tuning an embedding model can be less resource-intensive than retraining an entire LLM from scratch.

How to Fine-Tune Embedding Models

Fine-tuning typically involves adjusting the pretrained weights of the embedding layer (and sometimes the entire model) using a task-specific dataset. Here's a general workflow:

1. **Select a Pretrained Model**
   a. Start with a model like BERT, DistilBERT, or a sentence-transformer (e.g., all-MiniLM-L6-v2) suited to your task.
   b. Choose based on size, speed, and whether it's designed for sentence-level or token-level embeddings.
2. **Prepare a Task-Specific Dataset**
   a. Collect labeled data relevant to your task (e.g., positive/negative reviews for sentiment analysis).
   b. For unsupervised fine-tuning, use unlabeled domain-specific text (e.g., scientific papers) and a self-supervised objective like contrastive learning.
3. **Choose a Fine-Tuning Strategy**
   a. **Full Fine-Tuning**: Update all model parameters, including the embedding layer and subsequent layers. This is computationally expensive but often yields the best results.
   b. **Embedding-Only Fine-Tuning**: Adjust only the embedding layer while freezing the rest of the model. This is faster and useful when computational resources are limited.
   c. **Adapter-Based Fine-Tuning**: Add small, task-specific layers (adapters) to the model while keeping the original embeddings mostly frozen. This balances efficiency and performance.

4. **Define a Loss Function**

   a. For supervised tasks: Use cross-entropy loss (classification), mean squared error (regression), etc.

   b. For unsupervised tasks: Use contrastive loss (e.g., InfoNCE) or triplet loss to bring similar embeddings closer and push dissimilar ones apart.

5. **Train the Model**

   a. Use a framework like PyTorch or Hugging Face's Transformers library.

   b. Set hyperparameters: learning rate (e.g., 2e–5), batch size, and epochs. A smaller learning rate is often preferred to avoid catastrophic forgetting of pretrained knowledge.

   c. Monitor performance on a validation set to prevent overfitting.

6. **Evaluate and Iterate**

   a. Test the fine-tuned embeddings on your task (e.g., cosine similarity for semantic search, accuracy for classification).

   b. Adjust the dataset, loss function, or strategy if results are suboptimal.

Popular Techniques for Fine-Tuning Embeddings

- **Masked Language Modeling (MLM)**: Continue pretraining on domain-specific data by masking words and predicting them, refining the embeddings for that domain.

- **Contrastive Learning**: Train embeddings to distinguish positive pairs (similar texts) from negative pairs (dissimilar texts), common in sentence-transformers.

- **Prompt-Based Fine-Tuning**: Use task-specific prompts to guide the model, indirectly influencing embeddings without extensive retraining.

- **Knowledge Distillation**: Fine-tune a smaller embedding model by learning from a larger, pretrained LLM, preserving quality while reducing size.

c. Full Model Fine-Tuning

Full fine-tuning, on the other hand, takes a more comprehensive approach. Instead of freezing parts of the model, this method trains the entire model on task-specific data. Each layer of the model is updated, allowing it to adapt fully to the nuances of the new dataset.

This approach is especially advantageous when the task-specific dataset is large or significantly differs from the data used in pretraining. By enabling all layers to learn from the new data, full fine-tuning fosters a deeper and more precise alignment between the model and the task, often leading to superior performance. However, this process demands greater computational resources, more training time, and meticulous management to avoid overfitting or destabilizing the model.

Striking the Balance

Both approaches carry immense potential, and the choice between them depends on the complexity of the task, the availability of data, and the computational resources at hand. Whether repurposing the model's pretrained strengths through feature extraction or deeply reconfiguring it with full fine-tuning, these methods highlight the incredible adaptability of LLMs—bringing us closer to uncovering new possibilities and solutions in a world full of linguistic complexity.

Prominent Fine-Tuning Methods

Fine-tuning large language models (LLMs) involves adjusting their parameters to meet specific requirements. These methods are broadly categorized into **supervised fine-tuning** and **reinforcement learning from human feedback (RLHF)**, each offering distinct techniques to adapt models effectively to targeted applications. Below is an in-depth exploration of these methods.

a. Supervised Fine-Tuning

Supervised fine-tuning uses labeled datasets where each input is paired with a correct label or output. The model learns by adjusting its parameters to predict these labels accurately. This method builds on the model's pre-existing knowledge from pretraining, adapting it to specific tasks. It is widely used for customizing LLMs and improving task-specific performance.

Key Techniques in Supervised Fine-Tuning

1. **Basic Hyperparameter Tuning:** This straightforward approach involves manually adjusting hyperparameters like learning rate, batch size, and the number of epochs to optimize the model's performance. The goal is to balance learning efficiency with the risk of overfitting. Well-chosen hyperparameters enhance the model's ability to generalize and improve its accuracy on specific tasks.

2. **Transfer Learning** is ideal when task-specific data is limited. The model, pretrained on a large general dataset, is fine-tuned with a smaller, task-specific dataset. This method significantly reduces the training time and data requirements while often delivering superior results compared to training from scratch. It effectively repurposes the knowledge embedded in the model for new applications.

3. **Multitask Learning:** In this technique, the model is fine-tuned on multiple related tasks simultaneously. The shared learning process helps the model generalize better across tasks, leveraging common patterns and relationships. It is particularly beneficial when individual tasks have limited labeled data, as the combined dataset provides richer training signals. Multitask learning requires labeled datasets for each task and improves performance for closely related tasks.

4. **Few-Shot Learning** enables the model to perform a task with minimal labeled data. The model relies on its pre-existing knowledge from pretraining, using a few examples to adapt to the new task. During inference, prompts include examples or "shots" to guide the model's responses. This technique is highly effective when labeled data is scarce or expensive to obtain and can complement RLHF when human feedback is incorporated.

5. **Task-Specific Fine-Tuning** focuses entirely on optimizing the model for a particular task. This approach involves refining the model's parameters to suit the domain's unique nuances,

improving accuracy and relevance. While related to transfer learning, task-specific fine-tuning hones in on the exact requirements of a single task rather than broadly adapting pretrained features.

b. Reinforcement Learning from Human Feedback (RLHF)

RLHF is an innovative approach where human feedback is integrated into the model's training process. This method trains models to produce outputs that align with human expectations, leveraging human evaluators' expertise and judgment to improve contextual and practical accuracy.

Key Techniques in RLHF

1. **Reward Modeling** uses human evaluations to guide the model's learning. The model generates multiple outputs, which are ranked or scored by human evaluators. Based on these rankings, the model predicts the human-provided rewards and adjusts its behavior to maximize these rewards. This technique allows the model to learn complex tasks defined by nuanced human preferences.

2. **Proximal Policy Optimization (PPO)** is a reinforcement learning algorithm designed to optimize the model's policy while maintaining stability. The model updates its parameters iteratively to maximize the expected reward. A constraint ensures that updates are incremental, avoiding drastic changes that could destabilize the model. This balance between exploration and stability makes PPO an efficient and reliable reinforcement learning method.

3. **Comparative Ranking** focuses on relative quality rather than absolute evaluation.

   Human evaluators rank multiple outputs, allowing the model to learn from these comparative judgments. By analyzing ranked outputs, the model improves its ability to generate higher-quality responses. This method provides nuanced feedback, helping the model understand subtle differences in output quality.

4. **Preference Learning** is a specialized form of RLHF where human evaluators provide preferences between pairs of outputs. The model learns to align its behavior with human preferences, even when explicit numerical rewards are difficult to define. This approach captures complex, subjective judgments, enabling the model to perform tasks requiring humanlike decision-making.

5. **Parameter-Efficient Fine-Tuning (PEFT)**

   PEFT focuses on updating only a subset of the model's parameters. By modifying specific layers or adding task-specific components, this method reduces computational and storage demands. PEFT maintains performance comparable to full fine-tuning while being resource-efficient, making it a practical choice for many applications.

Choosing the Right Method

The choice between supervised fine-tuning and RLHF—and the techniques within each—depends on the task's complexity, data availability, and desired outcomes. Supervised fine-tuning is ideal for well-defined tasks with labeled datasets, while RLHF excels in scenarios requiring nuanced, context-driven outputs guided by human judgment. Together, these methods highlight the adaptability and potential of fine-tuning in harnessing the full power of LLMs.

Fine-Tuning Process and Best Practices

Fine-tuning a pretrained language model to meet specific use case requirements involves following a structured process. This ensures that the model is optimized to deliver accurate and effective results. Below are the key steps and best practices for fine-tuning, along with examples of its applications.

a. Data Preparation

Data preparation is a foundational step in the fine-tuning process. It involves curating and preprocessing the dataset to ensure relevance and quality for the target task. This step typically includes the following:

- **Cleaning the Data:** Removing duplicates, incomplete entries, or irrelevant information
- **Handling Missing Values:** Addressing gaps in the dataset to maintain consistency
- **Formatting the Text:** Aligning the data structure with the model's input requirements

Data augmentation techniques, such as paraphrasing or synonym replacement, can also expand the dataset and improve the model's robustness. Proper data preparation directly impacts the model's ability to learn and generalize effectively, leading to enhanced task-specific performance and accurate outputs.

b. Choosing the Right Pretrained Model

Selecting a pretrained model that aligns with the specific requirements of your task is crucial for successful fine-tuning. Key considerations include

- **Model Architecture:** Understanding the layers and configurations of the model
- **Input/Output Specifications:** Ensuring compatibility with the task
- **Model Size and Training Data:** Balancing computational resources and task requirements
- **Performance Benchmarks:** Reviewing how well the model performs on tasks similar to yours

Choosing a pretrained model that closely matches the target task helps streamline the fine-tuning process and maximize its adaptability and effectiveness in your application.

c. Identifying the Right Parameters for Fine-Tuning

Configuring fine-tuning parameters ensures optimal learning and adaptation to task-specific data. Key parameters include

- **Learning Rate:** Determines how quickly the model updates during training

- **Batch Size:** Influences the efficiency of gradient calculations
- **Number of Epochs:** Controls the training duration to avoid underfitting or overfitting

Freezing certain layers (typically the earlier ones) while training the later layers is a common practice. This helps retain general knowledge from pretraining while allowing the model to adapt to task-specific requirements. Striking a balance between leveraging pretrained knowledge and learning new features is key to effective fine-tuning.

d. Validation

Validation ensures the fine-tuned model performs as expected on unseen data. This step involves

- Using a **validation dataset** to evaluate performance
- Monitoring metrics such as **accuracy, loss, precision, and recall** to assess the model's generalization capabilities

Validation highlights areas where the model may need further improvement, enabling adjustments to parameters or data to optimize performance. Regular validation throughout the fine-tuning process ensures consistent alignment with task goals.

Evaluation Metrics and Benchmarks for Fine-Tuning LLMs

When fine-tuning large language models (LLMs), it's important to apply the right evaluation metrics and benchmarks to assess performance accurately. The choice of metric depends heavily on the task (e.g., classification, generation, reasoning, etc.).

Classification Tasks (e.g., Sentiment Analysis, Intent Detection)

- **Accuracy**: Measures the proportion of correct predictions
- **Precision/Recall/F1 Score**: Especially useful for imbalanced datasets
- **ROC-AUC**: Captures the model's ability to distinguish between classes
- **Confusion Matrix**: Offers insights into types of classification errors

Text Generation Tasks (e.g., Summarization, Translation, Dialogue)

- **BLEU**: Based on n-gram overlap; commonly used in translation
- **ROUGE**: Measures recall; widely used in summarization tasks
- **METEOR**: Accounts for synonymy and stemming
- **BERTScore**: Uses contextual embeddings to assess semantic similarity
- **GLEU/chrF**: Variants of BLEU that better capture fluency in certain cases

Reasoning and Question Answering

- **Exact Match (EM)**: Measures strict correctness of answers
- **F1 Score**: Based on token overlap between the predicted and reference answers
- **Accuracy@k/Hits@k**: Common in retrieval and multiple-choice settings
- **Faithfulness/Consistency**: Often assessed through human evaluation

Dialogue and Chatbot Evaluation

- **BLEU/METEOR/ROUGE**: Evaluate fluency and relevance
- **DialogRPT/USR**: Model-based metrics that approximate human judgments
- **Human Evaluation**: Often necessary to assess coherence, appropriateness, and personality

General Model Evaluation

- **Perplexity**: Reflects how well the model predicts text (lower is better)
- **Log-Likelihood**: Useful for comparing model variants
- **Toxicity/Bias Scores**: Measured with external tools or datasets, such as Perspective API or RealToxicityPrompts

Common Benchmarks

Language Understanding

- **GLUE/SuperGLUE**: A suite of diverse tasks including sentiment, entailment, and coreference
- **MMLU (Massive Multitask Language Understanding)**: Tests knowledge across a wide range of academic subjects
- **BBH (Big-Bench Hard)**: A challenging benchmark for reasoning
- **HellaSwag/WinoGrande**: Focused on commonsense and pronoun resolution

Summarization

- **CNN/DailyMail**, **XSum**, **Gigaword**: Used to evaluate abstractive summarization performance

Machine Translation

- **WMT**: A standard benchmark for translation tasks with yearly competitions

Question Answering

- **SQuAD**, **NaturalQuestions**, **TriviaQA**, **HotpotQA**: Range from fact-based to reasoning-heavy question answering

Dialogue

- **PersonaChat**, **DSTC**, **MultiWOZ**: Datasets for evaluating both open-domain and task-oriented dialogue systems

Retrieval and Retrieval-Augmented Generation (RAG)

- **BEIR**: A diverse benchmark suite for retrieval-based tasks
- **MS MARCO**: Commonly used for passage ranking and open-domain QA

Best Practices

- Use a combination of metrics for a more comprehensive evaluation.
- Include both automated and human evaluations, especially for subjective tasks.
- Track metric changes before and after fine-tuning to measure improvements.
- Consider using LLM-as-a-judge or prompt-based evaluation for complex outputs.

e. Detect Bias, Fairness, and Groundedness of LLMs

Detecting bias, fairness, and groundedness in large language models (LLMs) is a critical task, especially when evaluating retrieval-augmented generation (RAG) systems. Frameworks like RAGAS and TruLens provide structured approaches to assess these qualities using specific metrics and methodologies.

Groundedness

Groundedness measures how well an LLM's response is supported by the retrieved context or source material, ensuring it doesn't hallucinate or deviate from the provided information. Both RAGAS and TruLens offer ways to evaluate this.

- **RAGAS Framework**
    - **Metric**: *Faithfulness* is the primary metric for groundedness in RAGAS. It assesses whether the LLM's response aligns with the retrieved context by breaking the response into individual statements and verifying each against the source material.
    - **How It Works**: RAGAS uses an LLM to evaluate the response. For each statement in the output, it checks if the retrieved context supports it, often employing a chain-of-thought reasoning process to provide a score (e.g., 0 to 1) and explanations. A low faithfulness score indicates potential hallucinations or unsupported claims.
    - **Implementation**: You provide the query, retrieved context, and LLM-generated response. RAGAS then computes the faithfulness score by analyzing factual consistency.
- **TruLens Framework**
    - **Metric**: *Groundedness* is explicitly measured in TruLens as part of the RAG Triad (context relevance, groundedness, answer relevance). It evaluates how well each part of the response is anchored in the retrieved context.
    - **How It Works**: TruLens uses a feedback function powered by an LLM (e.g., GPT-3.5) to score groundedness. It parses the response into segments and checks their alignment with the context, providing a score and reasoning for transparency.
    - **Implementation**: Using TruLens, you set up an evaluator with a Tru object and a recorder to log the query, context, and response. The framework then runs the groundedness evaluation, allowing you to tweak parameters like chunk size or retrieval strategy based on results.

Bias

Bias in LLMs refers to unfair or skewed outputs that reflect prejudices in training data or model behavior, often related to demographics, ideologies, or social groups. While RAGAS and TruLens don't directly target bias as a standalone metric, their evaluation techniques can be adapted to detect it.

- **RAGAS Framework**
    - **Approach**: Bias isn't a predefined metric in RAGAS, but you can detect it indirectly through *faithfulness* and *answer relevance*. For example, if an LLM consistently generates responses that misrepresent certain groups (e.g., gender or race) despite accurate context, this could indicate bias.
    - **How to Detect**: Create a diverse set of queries and contexts targeting sensitive attributes (e.g., "Describe a typical software engineer" with contexts mentioning different genders). Compare the faithfulness scores across these responses. Disparities in how the LLM interprets or uses context for different groups may suggest bias.
    - **Limitations**: RAGAS focuses on factual alignment, so subtle biases (e.g., tone or omission) might require additional qualitative analysis or custom metrics.
- **TruLens Framework**
    - **Approach**: TruLens also lacks a direct bias metric but can be extended to assess bias through groundedness and answer relevance evaluations across varied inputs.
    - **How to Detect**: Test the LLM with prompts designed to probe for bias (e.g., "Provide a job recommendation for a male vs. female candidate" with identical contexts). Analyze the groundedness scores to see if the LLM deviates from the context differently based on demographic factors. Low groundedness for specific groups might indicate biased interpretation.

- **Customization**: TruLens allows custom feedback functions. You could define a bias-specific metric by comparing response patterns across demographic variations, leveraging its systematic experiment tracking to establish baselines.

Fairness

Fairness evaluates whether an LLM treats different groups equitably, avoiding discrimination or unequal performance. Neither RAGAS nor TruLens has an explicit fairness metric, but their evaluation pipelines can be adapted to assess fairness indirectly.

- **RAGAS Framework**
    - **Approach**: Use *context recall* and *answer relevance* to check if the LLM retrieves and uses context equitably across groups. Context recall measures how much of the relevant context is included, while answer relevance ensures the response addresses the query appropriately.
    - **How to Detect**: Design evaluation datasets with balanced representation (e.g., equal mentions of different ethnicities or genders in contexts). Run RAGAS to compute recall and relevance scores for each group. Significant score variations (e.g., higher relevance for one gender) could indicate unfairness in retrieval or generation.
    - **Practical Steps**: Generate synthetic datasets with counterfactuals (e.g., swapping gender in prompts), and analyze if the LLM's performance remains consistent.
- **TruLens Framework**
    - **Approach**: Leverage the RAG Triad to assess fairness by ensuring context relevance, groundedness, and answer relevance are consistent across diverse inputs.
    - **How to Detect**: Test the LLM with a dataset covering multiple demographic groups (e.g., FairFace or Bias in Bios). Evaluate the triad metrics for each group. For instance, if context relevance

is lower for underrepresented groups, it might suggest biased retrieval; if answer relevance varies, it could point to unfair generation.

- **Experimentation**: TruLens supports iterative testing. Adjust retrieval parameters (e.g., sentence window size) and observe their impact on fairness metrics, aiming for uniform performance across groups.

Practical Steps to Implement

1. **Dataset Preparation**
   a. Curate a diverse evaluation set with queries and contexts spanning demographics, ideologies, or other bias-prone areas.
   b. Include counterfactual examples (e.g., changing "he" to "she" in prompts) to test consistency.
2. **RAGAS Setup**
   a. Install RAGAS (pip install ragas), and input your query, context, and response.
   b. Run faithfulness and relevance evaluations, and then analyze scores for patterns indicating bias or unfairness.
3. **TruLens Setup**
   a. Install TruLens (pip install trulens-eval), and initialize a Tru object.
   b. Define a recorder with your RAG pipeline, and run evaluations using the RAG Triad. Compare results across groups.

RAGAS Example:

```
from ragas import evaluate
from datasets import Dataset

data = Dataset.from_dict({
    "question": ["What's France's capital?"],
    "context": ["France's capital is Paris."],
    "answer": ["Paris"]
```

```
})
result = evaluate(data, metrics=["faithfulness"])
print(result["faithfulness"])
Output: 1.0 (grounded)
```

If answer were “London,” score would be ~0.0 (ungrounded).

TruLens Example:

```
from trulens_eval import Tru, Feedback from trulens_eval.feedback import
Groundedness
tru = Tru()
groundedness = Groundedness()
feedback = Feedback(groundedness.groundedness_measure)
result = tru.run_feedback_functions( record={"query": "2020 election
winner?", "context": "Joe Biden won.", "response": "Joe Biden"} )
print(result)
Output: ~0.9 (grounded)
```

Detecting Data Drift When Fine-Tuning

Detecting data drift when fine-tuning a large language model (LLM) is crucial to ensure the model remains effective and generalizes well to new data. Data drift occurs when the distribution of the incoming data (e.g., the fine-tuning dataset or real-world inference data) diverges from the distribution of the original training dataset. Here’s a step-by-step approach to detect data drift during fine-tuning:

1. **Define Key Metrics and Features**

    - **Text Features:** Extract relevant features from your dataset, such as token frequency, sentence length, vocabulary size, n-gram distributions, or embeddings (e.g., from a pretrained model like BERT).

    - **Task-Specific Metrics:** If fine-tuning for a specific task (e.g., classification), monitor label distributions, class balance, or other task-relevant statistics.

    - **Baseline:** Use the original training dataset (or a representative subset) as a reference for comparison.

2. **Statistical Tests**

    - **Distribution Comparison:** Apply statistical tests to compare the original training data and the fine-tuning data:

        - **Kolmogorov-Smirnov (KS) Test:** For continuous features like sentence length or embedding distances

        - **Chi-Square Test:** For categorical data like label distributions or token frequencies

        - **Wasserstein Distance:** Measures the "distance" between two distributions, useful for embeddings or numerical features

    - **Thresholds:** Set significance thresholds (e.g., p-value < 0.05) to flag significant drift.

3. **Embedding-Based Drift Detection**

    - **Generate Embeddings:** Use the pretrained LLM (before fine-tuning) to encode both the original training data and the fine-tuning data into a latent space (e.g., mean-pooled embeddings).

    - **Compare Distributions:** Calculate drift using metrics like

        - **Cosine Similarity:** Between average embeddings of the two datasets

        - **Maximum Mean Discrepancy (MMD):** A kernel-based method to measure divergence between distributions

        - **KL Divergence:** If you can estimate probability densities (e.g., via histograms or kernel density estimation)

        - **Visualization:** Use t-SNE or PCA to visualize embeddings and spot clusters or shifts

4. **Monitor Model Performance**

    - **Validation Set:** Maintain a held-out validation set from the original training distribution. Track performance metrics (e.g., accuracy, perplexity, F1 score) during fine-tuning.

- **Performance Drop:** A significant drop might indicate the fine-tuning data is drifting too far from the original distribution, causing the model to overfit or lose generalization.
- **Cross-Dataset Evaluation:** Periodically evaluate the fine-tuned model on both the original validation set and a sample of the fine-tuning data to detect discrepancies.

5. **Concept Drift in Task-Specific Fine-Tuning**
   - **Label Shift:** Check if the label distribution changes (e.g., a sentiment model seeing more negative samples in fine-tuning than in training).
   - **Covariate Shift:** Compare input feature distributions (e.g., topics, vocabulary) while assuming the task remains the same.
   - **Semantic Shift:** Use topic modeling (e.g., LDA) or keyword analysis to detect changes in the underlying themes or concepts.
6. **Practical Example**
   - Suppose you're fine-tuning an LLM for customer support classification:
     - Extract token frequencies and embeddings from the original training data (e.g., product reviews) and the fine-tuning data (e.g., live chat logs).
     - Run a KS test on sentence lengths and a Wasserstein distance on embeddings.
     - If p-values indicate significant drift or distances exceed a threshold, investigate further (e.g., new slang in chats not present in reviews).
7. **Mitigation**

   If drift is detected, consider

   - **Reweighting:** Adjust the fine-tuning data to align with the original distribution.

- **Regularization:** Use techniques like weight decay or domain-adversarial training to reduce overfitting to drifted data.
- **Data Augmentation:** Blend original and fine-tuning data to smooth the transition.

f. Model Iteration

Iteration involves refining the model based on evaluation results. This step includes

- Adjusting fine-tuning parameters, such as learning rate or the extent of layer freezing
- Implementing regularization techniques to prevent overfitting
- Exploring alternative architectures or training strategies

Iterative improvements allow engineers to progressively enhance the model's capabilities, ensuring it meets the desired performance levels before deployment.

g. Model Deployment

Deployment transitions the fine-tuned model from development to real-world application. Key considerations during this phase include

- Ensuring **hardware and software compatibility** with the deployment environment
- Integrating the model into existing systems or workflows
- Addressing scalability, real-time performance, and security measures

Successful deployment ensures the model operates seamlessly in its intended environment, delivering the enhanced capabilities achieved through fine-tuning.

Fine-Tuning Applications

Fine-tuning pretrained models is a powerful way to adapt general-purpose LLMs for specific tasks. Below are some of the most prominent use cases where fine-tuning offers significant benefits.

a. Sentiment Analysis

Fine-tuned models enable accurate sentiment analysis, providing insights from customer feedback, social media posts, and product reviews. Businesses can use these insights to

- Identify trends and gauge customer satisfaction
- Inform marketing strategies and product development
- Track public sentiment for proactive reputation management

For example, a company might fine-tune a model on its specific customer data to better understand feedback nuances, helping drive targeted improvements and customer engagement.

b. Chatbots

Fine-tuning enhances chatbot performance, enabling more engaging and contextually relevant conversations. Applications include

- **Customer Service:** Providing personalized assistance and resolving queries
- **Healthcare:** Answering medical questions and offering patient support
- **Ecommerce:** Assisting with product recommendations and transactions
- **Finance:** Offering personalized financial advice and account management

By adapting language models to specific industries, fine-tuned chatbots become valuable tools for improving user interactions and customer satisfaction.

c. Summarization

Fine-tuned models can generate concise, informative summaries of lengthy documents, articles, or conversations, streamlining information retrieval. Applications include

- **Academic Research:** Condensing research papers for quick understanding
- **Corporate Environments:** Summarizing reports and emails to aid decision-making
- **Legal and Medical Fields:** Providing summaries of case files or patient histories for efficient review

Fine-tuned summarization models enable professionals to process vast amounts of information more effectively, improving productivity and knowledge management.

Fine-tuning pretrained language models unlocks their potential to deliver optimized, task-specific outcomes. By following a structured process and adhering to best practices in data preparation, parameter configuration, validation, iteration, and deployment, organizations can harness the power of LLMs to address unique challenges. From sentiment analysis and chatbots to summarization, fine-tuned models demonstrate versatility and effectiveness, offering significant benefits across industries and applications.

Advanced Fine-Tuning Techniques for LLMs

As large language models (LLMs) grow in size and complexity, traditional fine-tuning approaches can become computationally expensive, resource-intensive, or insufficient for specialized needs. Advanced fine-tuning techniques have emerged to address these limitations, offering innovative ways to adapt LLMs efficiently and effectively. This section explores four prominent methods—Low-Rank Adaptation (LoRA), Prompt Tuning, Continual Learning, and Federated Fine-Tuning—each pushing the boundaries of how LLMs can be customized for diverse applications.

Low-Rank Adaptation (LoRA)

Low-Rank Adaptation (LoRA) is a parameter-efficient fine-tuning technique that updates only a small subset of a model's weights, reducing the computational and memory burden of full fine-tuning. Instead of modifying all parameters, LoRA introduces low-rank updates to specific weight matrices (e.g., in the attention layers), allowing the model to adapt to new tasks while keeping the original pretrained weights frozen.

Mechanics

- LoRA assumes that the changes needed for task-specific adaptation lie in a low-dimensional subspace of the full weight matrix.
- For a weight matrix W W W in the model, LoRA adds a low-rank decomposition $\Delta W{=}A{\cdot}B$ *\Delta W = A \cdot B* ΔW=A·B, where A A A and B B B are smaller matrices with rank r r r (much smaller than the original dimensions).
- During fine-tuning, only A A A and B B B are trained, while W W W remains unchanged. The updated weights are computed as $W'{=}W{+}\Delta W$ *W' = W + \Delta W* W′=W+ΔW during inference.

Advantages

- **Efficiency**: Reduces memory usage and training time significantly (e.g., fine-tuning a billion-parameter model might require updating only 0.1% of parameters).
- **Modularity**: Task-specific updates can be stored separately and swapped in or out without altering the base model.
- **Scalability**: Ideal for fine-tuning massive models like GPT-3 or LLaMA on resource-constrained hardware.

Challenges

- May underperform full fine-tuning on highly specialized tasks requiring extensive adaptation
- Requires careful selection of the rank r r r to balance efficiency and expressiveness

Use Cases

- Fine-tuning LLMs for multiple domain-specific chatbots (e.g., legal, medical) with minimal storage overhead
- Adapting large models on edge devices where memory and compute are limited

Example: A company fine-tunes a 175-billion-parameter LLM for customer support using LoRA, reducing the trainable parameters from 175 billion to a few million, achieving comparable performance to full fine-tuning with a fraction of the GPU hours.

Prompt Tuning

Prompt Tuning shifts the focus from modifying model weights to optimizing task-specific prompts, leveraging the pretrained LLM's inherent capabilities without altering its parameters. This method is particularly useful for extremely large models where full fine-tuning is impractical.

Mechanics

- Instead of updating the model, a set of trainable prompt embeddings (virtual tokens) is prepended to the input sequence.
- These embeddings are optimized during training to guide the model toward desired outputs for a specific task.
- The pretrained weights remain frozen, and only the prompt embeddings (a tiny fraction of parameters) are adjusted.

Advantages

- **Ultraefficient**: Requires updating far fewer parameters than even PEFT methods like LoRA (e.g., tens of thousands vs. millions).
- **Preserves Model Integrity**: Avoids risks of overfitting or catastrophic forgetting since the core model is unchanged.
- **Flexibility**: Prompts can be easily swapped for different tasks, making it ideal for multitask scenarios.

Challenges

- Performance may lag behind full fine-tuning for complex tasks requiring deep adaptation.
- Designing effective initial prompts can be nontrivial and task-dependent.

Use Cases

- Rapid prototyping of task-specific applications (e.g., sentiment analysis, summarization) without retraining the model
- Deploying a single LLM to handle multiple tasks by switching prompts dynamically (e.g., a virtual assistant toggling between scheduling and translation)

Example: An ecommerce platform uses prompt tuning to adapt a pretrained LLM for product description generation, training a 100-token prompt to produce concise, brand-aligned outputs without touching the model's 70B parameters.

Federated Fine-Tuning

Federated Fine-Tuning takes fine-tuning into decentralized territory, training LLMs across multiple devices or institutions without centralizing sensitive data, a critical feature for privacy-sensitive fields like healthcare or finance. In this setup, local models are fine-tuned on individual datasets—say, patient records at different hospitals—and their updates are aggregated into a global model without ever sharing the raw data. This aggregation typically uses techniques like federated averaging, where weight updates are combined to refine the shared model.

The result is a collaboratively trained LLM that respects data privacy and complies with regulations like GDPR or HIPAA, all while leveraging diverse datasets. However, this approach faces hurdles: coordinating training across heterogeneous devices can be complex, and differences in data distribution may lead to suboptimal performance. A consortium of hospitals might employ Federated Fine-Tuning to develop a diagnostic chatbot, each contributing local insights to a shared model without compromising patient confidentiality.

Together, these advanced techniques highlight the evolving landscape of LLM fine-tuning, offering solutions to the practical and ethical challenges of adapting massive models. LoRA and Prompt Tuning excel in efficiency, making fine-tuning accessible even for resource-constrained settings, while Continual Learning ensures models remain versatile over time. Federated Fine-Tuning, meanwhile, bridges the gap between customization and privacy, opening doors to collaborative AI development.

Each method carries unique strengths and trade-offs, and their application depends on the task, resources, and constraints at hand. By mastering these approaches, practitioners can unlock the full potential of LLMs, tailoring them to an ever-widening array of real-world challenges with precision and responsibility.

When to Not Use LLM Fine-Tuning

Fine-tuning large language models (LLMs) has revolutionized how AI can be tailored to specific tasks and domains, but it is not always the best or most appropriate approach. In some situations, fine-tuning might not provide a clear advantage, and in others, it may even introduce risks or inefficiencies. Understanding the limitations and trade-offs of fine-tuning is crucial for making informed decisions about whether it is the right approach for your use case. Below is an in-depth exploration of when and why fine-tuning may not be suitable.

Pretrained Models Are Already Sufficient

Pretrained LLMs, like GPT and similar models, are designed to handle a broad range of language tasks effectively. They have been trained on massive datasets covering diverse topics, allowing them to perform well in many general-purpose scenarios without additional fine-tuning. For instance, tasks like summarization, basic question answering, and translation often yield satisfactory results using pretrained models. By leveraging prompt engineering, users can guide the model to perform specific tasks by simply designing inputs that include instructions or examples.

For example, a customer service application might ask the model to generate polite responses to common questions. By crafting a few-shot prompt with sample questions and answers, the pretrained model can adapt its output to align with the desired tone and style. This avoids the need for fine-tuning, which would involve additional costs and complexity. Fine-tuning in such cases may only yield marginal improvements, making it an inefficient use of resources.

Insufficient or Low-Quality Data

Fine-tuning requires access to task-specific data that is not only sufficient in quantity but also high in quality. The dataset should be clean, well-labeled, and representative of the domain the model will be applied to. When these criteria are not met, fine-tuning can introduce significant challenges.

If the dataset is too small, the model risks overfitting to the limited examples, which could lead to poor generalization to new inputs. For example, if a legal document analysis model is fine-tuned on only a handful of annotated cases, it might perform well on similar examples but fail when presented with novel or slightly different legal contexts. Moreover, if the dataset contains errors, inconsistencies, or biases, the model might incorporate these issues into its outputs, amplifying them in unintended ways.

In cases where high-quality data is unavailable or difficult to curate, other approaches, such as few-shot learning, transfer learning, or prompt engineering, may be more practical. These methods allow the model to perform tasks effectively without relying heavily on extensive task-specific datasets.

High Computational Costs and Resource Constraints

Fine-tuning LLMs can be resource-intensive, requiring significant computational power, time, and storage. Training a large model, especially those with billions of parameters, involves running complex computations across high-performance hardware like GPUs or TPUs. This can result in prohibitive costs, particularly for organizations with limited budgets or infrastructure.

The fine-tuned model may also demand additional storage and memory for deployment, especially if the updated parameters increase the overall size of the model. For lightweight applications or environments with strict resource constraints, such as mobile devices or edge computing, deploying a fine-tuned model may be impractical. Instead, relying on pretrained models as-is, or applying techniques like parameter-efficient fine-tuning (PEFT), can help achieve acceptable performance without the overhead of full fine-tuning.

Regulatory, Privacy, and Ethical Constraints

Certain industries, such as healthcare, finance, and government, are subject to stringent regulations around data privacy, security, and usage. Fine-tuning often involves training a model on proprietary or sensitive data, which can raise significant legal and ethical concerns. For example, fine-tuning a medical diagnostic model using patient records might violate data privacy regulations like HIPAA (Health Insurance Portability and Accountability Act) or GDPR (General Data Protection Regulation).

Even if data anonymization techniques are employed, there is always a risk that sensitive information could be inadvertently encoded in the model's parameters. This could lead to unintended exposure of confidential information, especially in scenarios where the model is accessed by third parties. In such cases, organizations might consider using techniques like reinforcement learning from human feedback (RLHF) or synthetic data generation to achieve their goals without compromising privacy.

Ethical concerns also arise when fine-tuning is performed without careful consideration of biases in the training data. If the dataset reflects societal biases or discriminatory practices, the fine-tuned model may perpetuate or amplify these biases, leading to harmful or unfair outcomes. Organizations must weigh these risks carefully and explore alternative methods that minimize ethical liabilities.

Maintaining Model Versatility

Fine-tuning customizes a model for a specific task, often at the expense of its general-purpose capabilities. For applications that require flexibility across multiple tasks or domains, this specialization can become a limitation. For instance, a model fine-tuned for legal text summarization might lose its ability to perform other tasks, such as conversational AI or financial analysis, as effectively as it did in its pretrained state.

This loss of versatility is particularly concerning in use cases where the model needs to operate in diverse contexts or adapt to evolving requirements. In such situations, techniques like adapter layers, which allow task-specific customization without altering the core model, or dynamic prompt engineering, which leverages the model's pretrained knowledge, can offer better solutions. These approaches preserve the model's general-purpose utility while enabling targeted improvements.

Task Scope Is Uncertain or Evolving

When the exact requirements of a task are unclear or likely to change over time, fine-tuning can become a costly and time-consuming iterative process. For example, an organization exploring AI applications in customer service might initially require a model to answer basic inquiries but later expand its scope to include complex problem-solving or multilingual support. Fine-tuning the model for each incremental change would be inefficient, requiring repeated adjustments to data, training processes, and deployment strategies.

In such exploratory contexts, pretrained models with flexible prompting capabilities are often a better choice. They allow for rapid prototyping and experimentation without the need for extensive fine-tuning. Once the requirements stabilize, organizations can evaluate whether fine-tuning or another optimization method is necessary.

High-Risk Scenarios Requiring Predictability and Stability

In high-stakes applications, such as legal decision-making, medical diagnoses, or financial forecasting, the predictability and stability of the model's behavior are paramount. Fine-tuned models, especially those trained on narrowly defined datasets, can exhibit unpredictable performance when encountering out-of-distribution inputs. This variability poses significant risks in scenarios where incorrect or unreliable outputs could have serious consequences.

For these applications, it may be better to rely on the more generalized capabilities of pretrained models, which are often more robust across a wider range of inputs. Additionally, employing methods like human-in-the-loop systems, where model outputs are reviewed and verified by domain experts, can enhance reliability without the need for fine-tuning.

While fine-tuning offers powerful customization options for large language models, it is not always the most appropriate or effective approach. Scenarios where the pretrained model already performs well, where data quality or quantity is insufficient, or where computational resources are limited make fine-tuning less viable. Similarly, regulatory and ethical concerns, the need for model versatility, uncertain task requirements, or high-risk applications may favor alternative strategies.

Organizations should carefully assess their goals, constraints, and the specific needs of their applications before deciding to fine-tune an LLM. By leveraging pretrained models through prompt engineering, few-shot learning, or lightweight customization techniques, many of the advantages of LLMs can be realized without the added complexities and risks associated with full fine-tuning. This thoughtful approach ensures efficient use of resources while maximizing the impact and effectiveness of AI solutions.

Ethics and Bias in AI and LLMs

The ethics of artificial intelligence (AI), particularly in the context of large language models (LLMs), is a cornerstone of responsible development and deployment. As LLMs become increasingly integrated into everyday applications, addressing their ethical dimensions is essential to ensure alignment with human values, societal well-being, and fundamental rights. This discussion explores the multifaceted nature of AI ethics, highlights specific challenges associated with LLMs, and examines actionable solutions to foster responsible AI development.

Understanding AI Ethics and Its Relevance to LLMs

AI ethics encompasses a set of principles, values, and guidelines aimed at ensuring that AI systems are designed and utilized responsibly. The ethical landscape for LLMs is particularly complex due to their linguistic nature and widespread applicability. These models influence communication, information dissemination, decision-making, and even creative processes, making their ethical alignment a critical priority.

Ethical considerations for LLMs include transparency, fairness, accountability, privacy, human agency, and societal impact. Unlike conventional AI systems, LLMs directly interface with human language, amplifying their potential to shape opinions, reinforce biases, and impact decision-making processes. The ethical challenges they present demand proactive engagement from researchers, developers, ethicists, policymakers, and society.

Core Ethical Challenges in LLMs

Bias in Language Models

Bias is one of the most pressing ethical concerns in LLMs. These models learn from vast datasets, which often reflect societal prejudices and inequalities. Consequently, LLMs can perpetuate or amplify biases in their outputs.

- **Types of Bias**

  Bias in LLMs manifests in various forms:

  - **Stereotypical Bias:** Reinforcing societal stereotypes related to race, gender, or ethnicity
  - **Gender Bias:** Unequal representation or treatment of genders in generated content
  - **Cultural Bias:** Misrepresentation or oversimplification of cultural nuances
  - **Political Bias:** Favoring certain political ideologies, potentially compromising neutrality

- **Sources of Bias**

  The primary sources of bias include

  - **Training Data:** The datasets used to train LLMs often contain historical inequalities and unbalanced representation.
  - **Algorithmic Bias:** The mathematical frameworks and optimization techniques can inadvertently introduce or amplify biases.

- **Impact of Bias**

  Bias in LLMs can result in discriminatory outputs, spread misinformation, and reinforce systemic inequalities. For instance, biased hiring systems or legal decision-making tools can perpetuate unfair practices, while misinformation in media amplifies distorted narratives.

Privacy and Data Usage

The datasets used to train LLMs often include text scraped from publicly available sources, raising concerns about privacy and data ownership. Training data may inadvertently contain sensitive personal information, leading to potential privacy breaches.

LLMs can also generate outputs that inadvertently reveal private or sensitive information. This challenge underscores the need for robust anonymization techniques, responsible data collection practices, and adherence to privacy laws such as GDPR and HIPAA.

Transparency and Accountability

LLMs operate as "black boxes," making it difficult to trace how specific outputs are generated. This lack of transparency poses challenges in understanding and auditing decision-making processes, particularly in high-stakes applications like healthcare, law, or finance. When errors or biased outputs occur, it becomes challenging to attribute responsibility, complicating accountability.

Misinformation and Manipulation

The ability of LLMs to generate realistic and humanlike text raises significant risks of misuse. They can be exploited to create fake news, spam, phishing content, or deep fakes, undermining trust in digital information ecosystems. Their role in amplifying misinformation makes it imperative to develop safeguards against malicious use.

Environmental Impact

The computational demands of training and running LLMs contribute to substantial energy consumption and carbon emissions. The environmental footprint of large-scale AI systems raises concerns about sustainability and aligns with broader societal goals to combat climate change.

Promoting Fairness and Equity in LLMs

Ensuring fairness and equity in LLMs involves addressing biases while fostering inclusivity. Achieving these goals requires targeted strategies:

1. **Diverse Training Data**

   Curating balanced and representative datasets reduces bias and ensures equitable representation of all groups.

2. **Fairness Metrics**

   Defining measurable fairness criteria provides benchmarks to assess and mitigate bias.

3. **Bias Auditing and Mitigation**

   Regular audits of LLM outputs help identify biased patterns. Techniques such as adversarial training and debiasing algorithms can mitigate identified biases.

4. **Human-Centered Design**

   Involving diverse stakeholders, including ethicists and domain experts, ensures the inclusion of varied perspectives in AI design and deployment.

Addressing Broader Ethical Concerns

Responsible AI Development

Responsible AI development demands a commitment to ethical principles:

- **Beneficence:** AI systems should prioritize societal well-being and avoid harm.
- **Transparency:** Clear documentation of training data, methodologies, and limitations is essential.
- **Accountability:** Developers must take responsibility for their systems' outputs and impacts.
- **Privacy:** Respecting individual privacy rights is nonnegotiable.

Embedding these principles into every stage of LLM development helps align their capabilities with ethical standards.

Regulation and Policy for Ethical AI

Effective governance frameworks are essential to address the ethical challenges posed by LLMs. Current efforts include

- **Transparency Reporting:** Mandating disclosure of data sources, methodologies, and known limitations
- **Ethics Review Boards:** Establishing independent review bodies to assess the societal implications of AI systems
- **Regulatory Compliance:** Enforcing adherence to data protection laws and ethical guidelines

Policy recommendations should focus on fostering collaboration between governments, industry leaders, and ethicists to establish standards for ethical AI development.

Future Directions

Advancing ethical practices in LLMs requires ongoing research and innovation. Key areas of focus include

- **Interpretable AI:** Enhancing the transparency of LLM decision-making processes
- **Energy Efficiency:** Developing greener algorithms and hardware to reduce environmental impact
- **Holistic AI Design:** Encouraging interdisciplinary collaboration to create culturally sensitive and ethical AI systems

Ethical considerations are integral to the responsible development and deployment of LLMs. By addressing issues such as bias, privacy, transparency, and environmental impact, the AI community can ensure that these technologies serve as tools for societal progress rather than harm. Through collaboration, regulation, and continuous innovation, LLMs can be aligned with human values, fostering trust, fairness, and accountability in their applications.

LLM Fine-Tuning Example

This example demonstrates fine-tuning GPT-2 for sentiment classification using the "mteb/tweet_sentiment_extraction" dataset. The process includes preparing the dataset, tokenizing the text, modifying GPT-2 with a classification head, and training the model to predict sentiment labels (positive, negative, neutral). By leveraging the pretrained capabilities of GPT-2, fine-tuning ensures efficient training, requiring less labeled data while achieving task-specific accuracy.

First, install the following libraries:

```
pip install datasets
pip install transformers
pip install evaluate
```

Step 1: Loading Dataset

```
dataset = load_dataset("mteb/tweet_sentiment_extraction")
```

This dataset is specifically designed for sentiment classification and contains

- **Text Data:** The tweets themselves, stored in the "text" column
- **Labels:** Sentiment annotations (e.g., positive, negative, neutral), stored in the "label" column

Fine-tuning requires a labeled dataset because the model learns to map inputs (tweets) to outputs (sentiment labels). The dataset is already split into training and testing subsets:

- **Training Set**: Used to adjust the model's weights during learning
- **Testing Set**: Used to evaluate the model's performance on unseen data

Step 2: Tokenization

Before the text can be fed into the model, it must be tokenized. Tokenization converts raw text into numerical representations (tokens) that the model can process:

```
tokenizer = GPT2Tokenizer.from_pretrained("gpt2")
tokenizer.pad_token = tokenizer.eos_token
```

The GPT-2 tokenizer maps each word, subword, or character in the text to an index in the model's vocabulary. For example, "Hello world!" might become [15496, 995, 0]. GPT-2 doesn't have a predefined padding token because it was designed for text generation tasks. For classification tasks, where inputs are batched together, all sequences must be the same length.

Padding ensures that shorter sequences are extended to match the longest sequence in a batch. Since GPT-2 lacks a specific padding token, its End-of-Sequence (EOS) token (<|endoftext|>) is used as a placeholder.

The tokenization function is defined as follows:

```
def tokenize_function(examples):
    return tokenizer(examples["text"], padding="max_length",
    truncation=True)
```

This function tokenizes the "text" column in the dataset while

- **Padding:** Ensuring all sequences in a batch have the same length by adding the padding token where necessary
- **Truncation:** Cutting off longer sequences that exceed the model's maximum input size (1024 tokens for GPT-2)

The dataset is tokenized using

```
tokenized_datasets = dataset.map(tokenize_function, batched=True)
```

This prepares the data for fine-tuning, converting raw text into numerical inputs compatible with GPT-2.

Step 3: Training and Evaluation Sets

To speed up training and experimentation, the code creates smaller subsets of the training and testing datasets:

```
small_train_dataset = tokenized_datasets["train"].shuffle(seed=42).
select(range(1000))
small_eval_dataset = tokenized_datasets["test"].shuffle(seed=42).
select(range(1000))
```

Note **Only 1000 examples are selected from each split.** This reduces computational load during development while retaining enough data to meaningfully fine-tune and evaluate the model.

Step 4: Adapting the Model

GPT-2, by default, is a generative model. To make it suitable for classification, it is adapted as follows:

```
model = GPT2ForSequenceClassification.from_pretrained("gpt2", num_labels=3)
model.config.pad_token_id = tokenizer.pad_token_id
```

- **GPT2ForSequenceClassification:** This class extends GPT-2 by adding a classification head—a linear layer that maps the model's outputs to a fixed number of labels (in this case, three sentiment classes: positive, negative, and neutral).
- **Retaining Pretrained Weights:** The pretrained weights in GPT-2's transformer layers are retained. These layers encode general language understanding, such as syntax and semantics. Fine-tuning updates these weights slightly to make the model focus on the nuances of sentiment analysis.
- **Padding Token ID:** The model is configured to recognize the padding token added during tokenization. This ensures the model ignores padding tokens during training and evaluation.

Step 5: Fine-Tuning the Model

The Trainer class simplifies the fine-tuning process by managing the training loop, including batching, gradient updates, and evaluation. Training is configured as follows:

```
training_args = TrainingArguments(
    output_dir="test_trainer",
    evaluation_strategy="epoch",
    per_device_train_batch_size=4,
```

```
    per_device_eval_batch_size=4,
    gradient_accumulation_steps=4,
    num_train_epochs=3,
    save_steps=1000,
    logging_dir="./logs",
    logging_steps=500,
)
```

- **Batch Size:** Determines how many examples are processed simultaneously. A smaller batch size reduces memory usage.
- **Gradient Accumulation:** Combines gradients over multiple batches before updating model weights. This effectively increases the batch size without exceeding memory limits.
- **Evaluation Strategy:** The model is evaluated at the end of each epoch.
- **Number of Epochs:** The training loop runs three times through the entire training set.

The Trainer is initialized with the following:

```
trainer = Trainer(
    model=model,
    args=training_args,
    train_dataset=small_train_dataset,
    eval_dataset=small_eval_dataset,
    tokenizer=tokenizer,
    compute_metrics=compute_metrics,
)
```

The Trainer

- Processes the training data in batches
- Computes the loss for each batch by comparing the model's predictions to the true labels

- Propagates the loss backward to compute gradients
- Updates the model's weights using an optimizer, gradually improving its ability to classify sentiment

Step 6: Evaluation

Once training is complete, the model's performance is evaluated on the test set:

```
results = trainer.evaluate()
print("Evaluation Results:", results)
```

During evaluation:

- The model processes unseen examples from the test set and predicts sentiment labels.
- Predictions are compared to the true labels, and the accuracy metric is computed to measure performance.

The compute_metrics function is defined to calculate accuracy:

```
def compute_metrics(eval_pred):
    logits, labels = eval_pred
    predictions = np.argmax(logits, axis=-1)
    return metric.compute(predictions=predictions, references=labels)
```

This function converts the model's raw predictions (logits) into class probabilities and calculates how many predictions match the true labels.

What Happens Internally During Fine-Tuning

1. **Forward Pass**
    - The input (tokenized tweets) passes through GPT-2's transformer layers. These layers process the input to produce contextualized representations for each token.

2. **Classification Head**

    - The classification head processes the output of the transformer layers, mapping the contextualized representations to the three sentiment classes (positive, negative, neutral).

3. **Loss Calculation**

    - The predicted sentiment logits are compared to the true labels using a loss function (e.g., cross-entropy loss). This quantifies how far off the predictions are.

4. **Backward Pass**

    - Gradients are computed by propagating the loss backward through the model. These gradients indicate how much to adjust each weight to reduce the loss.

5. **Weight Updates**

    - The optimizer updates the model's weights, gradually improving its ability to classify sentiment accurately.

By the end of fine-tuning:

The model becomes specialized for sentiment classification while retaining its general language understanding capabilities. The pretrained layers are slightly adjusted to focus on sentiment-related patterns in text. The classification head learns to map GPT-2's outputs to the sentiment labels effectively.

The whole code:

```
# Importing required libraries
from datasets import load_dataset
import pandas as pd
import numpy as np
from transformers import GPT2Tokenizer, GPT2ForSequenceClassification,
TrainingArguments, Trainer
import evaluate

# Loading the dataset
dataset = load_dataset("mteb/tweet_sentiment_extraction")
```

```
# Loading the tokenizer
tokenizer = GPT2Tokenizer.from_pretrained("gpt2")

# Setting the padding token
tokenizer.pad_token = tokenizer.eos_token

# Tokenization function
def tokenize_function(examples):
    return tokenizer(examples["text"], padding="max_length",
    truncation=True)

# Tokenizing the dataset
tokenized_datasets = dataset.map(tokenize_function, batched=True)

# Splitting the dataset into a smaller train and evaluation set
small_train_dataset = tokenized_datasets["train"].shuffle(seed=42).
select(range(1000))
small_eval_dataset = tokenized_datasets["test"].shuffle(seed=42).
select(range(1000))

# Loading the model
model = GPT2ForSequenceClassification.from_pretrained("gpt2", num_labels=3)

# Ensuring the model uses the same padding token
model.config.pad_token_id = tokenizer.pad_token_id

# Defining the evaluation metric
metric = evaluate.load("accuracy")

def compute_metrics(eval_pred):
    logits, labels = eval_pred
    predictions = np.argmax(logits, axis=-1)
    return metric.compute(predictions=predictions, references=labels)

# Defining training arguments
training_args = TrainingArguments(
    output_dir="test_trainer",
    evaluation_strategy="epoch",
    per_device_train_batch_size=4,  # Adjust batch size for your
    GPU/CPU memory
```

```
    per_device_eval_batch_size=4,
    gradient_accumulation_steps=4,  # For gradient accumulation
    num_train_epochs=3,
    save_steps=1000,
    logging_dir="./logs",
    logging_steps=500,
)

# Initializing the Trainer
trainer = Trainer(
    model=model,
    args=training_args,
    train_dataset=small_train_dataset,
    eval_dataset=small_eval_dataset,
    tokenizer=tokenizer,  # Ensure tokenizer is passed to Trainer
    compute_metrics=compute_metrics,
)

# Training the model
trainer.train()

# Evaluating the model
results = trainer.evaluate()
print("Evaluation Results:", results)
```

Output:

```
Evaluation Results: {'eval_loss': 0.8756747841835022, 'eval_accuracy':
0.724, 'eval_runtime': 104.1054, 'eval_samples_per_second': 14.408, 'eval_
steps_per_second': 3.602, 'epoch': 4.949333333333334}
```

Conclusion

Fine-tuning large language models represents a pivotal step in adapting general-purpose AI systems to meet the nuanced demands of real-world applications. This chapter has provided a comprehensive overview of the architectural foundations of LLMs, the strategies for customizing them through fine-tuning, and the evaluation frameworks necessary to ensure their effectiveness and reliability. From selecting the appropriate model architecture to implementing advanced techniques like LoRA, prompt tuning, and federated learning, practitioners are equipped with a diverse toolkit to enhance LLM performance across domains.

As the AI landscape continues to evolve, fine-tuning is not only a means of optimization—it is a practice that must be approached with rigor, responsibility, and adaptability. Ethical considerations, data quality, regulatory compliance, and resource constraints all play a critical role in determining whether fine-tuning is appropriate and how it should be executed. Evaluation metrics and benchmarks, including both automated and human-in-the-loop methods, further ensure that fine-tuned models align with intended goals while minimizing risk.

Ultimately, the ability to tailor LLMs for specific tasks, industries, or user needs is what transforms these models from powerful generalists into specialized, high-impact tools. Whether improving sentiment analysis, powering intelligent chatbots, or enabling domain-specific summarization, fine-tuning unlocks the full potential of large language models. The knowledge and strategies explored in this chapter lay the foundation for responsible and effective deployment of LLMs in today's data-driven world.

Index

D. Grigorov, *Intermediate Python and Large Language Models*, https://doi.org/10.1007/979-8-8688-1475-4

B

C

G, H

R, S

T, U, V

W, X

Y, Z

GPSR Compliance
The European Union's (EU) General Product Safety Regulation (GPSR) is a set of rules that requires consumer products to be safe and our obligations to ensure this.

If you have any concerns about our products, you can contact us on

ProductSafety@springernature.com

In case Publisher is established outside the EU, the EU authorized representative is:

Springer Nature Customer Service Center GmbH
Europaplatz 3
69115 Heidelberg, Germany

www.ingramcontent.com/pod-product-compliance
Lightning Source LLC
LaVergne TN
LVHW081332050326
832903LV00024B/1129
* 9 7 9 8 8 6 8 8 1 4 7 4 7 *